Indigenous Communication in Africa

Indigenous Communication in Africa: Concept, Applications and Prospects

Edited by
Kwasi Ansu-Kyeremeh
Director, School of Communication Studies University of Ghana, Legon

GHANA UNIVERSITIES PRESS
ACCRA
2005

Ghana Universities Press
P. O. Box GP 4219
Accra
Tel: 233(21) 513401, 513404
Fax: 233(21) 513402
E-mail: ghanauniversitiespress@yahoo.com

Distributed in Europe and North America
by African Books Collective
The Jam Factory, 27 Park End Street
Oxford OX1 1HU, UK
E-mail: abc@dial.pipex.com
Website: www.africanbookscollective.com

First published in 1998 in two volumes by the School of Communication
Studies, University of Ghana, Legon.

PRODUCED IN GHANA
Typesetting by Ghana Universities Press
Printed by Yamens Press Limited

To
PAUL A.V. ANSAH,
DEVOTE NGABIRANO
and
KWABENA N. BAME

CONTENTS

NOTES ON CONTRIBUTORS

Paul A.V. Ansah (*late*) was Professor and Director of the School of Communication Studies, University of Ghana. After his initial undergraduate studies in Ghana, he proceeded to the University of London where he obtained his M.A. and Ph.D. He later studied for another M.A. in communication in the University of Wisconsin from where he joined the then School of Journalism and Mass Communication at the University of Ghana. He had over 36 publications including journal articles and chapters of books. In 1986, his series of lectures during the Golden Jubilee celebrations of the Ghana Broadcasting Corporation were published under the title *Broadcasting and National Development*.

Njoku E. Awa is an Associate Professor in the Department of Communication at the College of Agriculture and Life Sciences, Cornell University. He obtained his B.A. and M.A. degrees in communication from the Michigan State University and his Ph.D. from Cornell University. His publications include several chapters in books and journal articles.

Kwabena N. Bame (*late*), a former senior education officer with the Ghana Ministry of Education and editor of the *Ghana Teachers Journal* (1963–65) was Associate Professor of Sociology in the Institute of African Studies, University of Ghana. After initial Bachelors and Masters degrees in Sociology from the University of Ghana, he studied for a B.Ed. and later Ph.D. from the University of Toronto. He was a visiting scholar with several institutions including University of Sokoto in Nigeria, Anderson College, South Carolina, Fulbright Scholar at Saint Augustine's College (Raleigh), North Carolina and Cornell University. His publications include three books and numerous articles. One of his books, *Come to Laugh: A Study of African Traditional Theatre in Ghana* won the Ghana Association of Writers George Padmore Award for the best work in Sociology, Fine Arts and Culture in 1986.

Louise M. Bourgault has taught mass communication in the Department of Communication and Performance Studies, Northern Michigan University, since 1984. She is a Professor of Mass Communication and holds a Ph.D. from Ohio University (Athens). Her research interests include development communication, press freedom, African mass media and the training of Third World media professionals. She has worked with African media professionals in twelve different countries and has lived in Africa for over seven years.

Kees P. Epskamp joined the Centre for the Study of Education in Developing Countries (CESO) immediately after obtaining his Masters in Social Anthropology in 1977. He is in charge of documentation and editor of publications. His research activities include evaluation of media-supported educational projects in South America and Africa. He also conducted extensive literature survey on the use of media, especially small media and theatre, for development. He is the Co-ordinator of the CESO inter-university course on Education in Developing Countries, and for several years guest lecturer at the Department of Theatre Studies of The University of Amsterdam. He has written many articles on development support communication, in particular on popular theatre. In 1989, he obtained a Ph.D. in Political and Cultural Sciences at the University of Amsterdam, subsequently publishing his dissertation, entitled *Theatre in Search of Social Change* at CESO.

Joy F. Morrison is an Assistant Professor of Mass Communications at the University of Alaska at Fairbanks. She also works as a consultant conducting KAP studies and research, and designing mass media campaigns for social marketing projects in Third World countries. She received her Ph.D. in Mass Communications at the University of Iowa in 1991. Her dissertation was based on a case study of forum theatre communication and social change in Burkina Faso.

Helen Nabasuta Mugambi, recently a Research Fellow at the Institute for Advanced Study and Research in the African Humanities, Northwestern University, is now an Assistant Professor in English and Comparative Literature at the State University of California at Fullerton. She obtained her M.A. and Ph.D. in comparative literature from Indiana University after her initial B.A. and Graduate Diploma in Mass Communication from Makerere University in Uganda.

Devote Ngabirano (*late*) At the time Devote passed away in 1993, she was enrolled in the Ph.D. program at The School of Telecommunication Studies, Ohio University.

Marie Riley is an asistant professor in the Department of Public Relations at Mount Saint Vincent University in Halifax, Nova Scotia. She taught at the School of Journalism and Mass Communication at the University of Ghana in the late 1970s and was affiliated to its succeeding School of Communication Studies between 1989 and 1990 during her doctoral field work. She holds a Ph.D. from Simon Fraser University, where she completed a doctoral thesis on indigenous forms of communication in Ghana. She has published in journals and contributed a chapter on communication and health care in Ghana to the forthcoming book: *Communication and the Transformation of Society: A Developing Region's Perspective*.

Des Wilson is a Senior Lecturer and acting head of the Department of Communication Arts, University of Uyo in Nigeria. He holds a Ph.D. in communication and language arts from the University of Ibadan in Nigeria. His doctoral thesis examined the contextual significance of indigenous communication in Nigeria. Specializing in ethnocommunicology, he has published widely in journals and contributed chapters to books both inside and outside Africa.

PREFACE

The subject of indigenous forms of communication may not seem that significant in an age of complex, fast and sophisticated technologically mediated communication systems. To those who have experienced their nature and the purposes they serve, however, the indigenous modes are still useful and invaluable communication assets in the societies within which they exist. This book, which is a compendium of various perspectives on indigenous sociocultural characteristics and how they impact on communication processes, is the result of the combined effort of scholars with research and applied experiences with the African cultural milieu. They document the structures and processes of the indigenous African communication modes against various theoretical backgrounds.

All the authors have firsthand experience with the situations about which they write as researchers, instructors, students, or field workers.

The research is mostly original, using primary sources and material. The differences are in the disciplinary approaches and methodology. This is another strength of the book because it affords the reader the variety that the holistic organizational structures of societies across the continent are characterized by.

Primarily, the material was assembled for reference purposes, to aid scholarly investigations into the phenomenon, and to provide basic information and bibliographical references which collectively serve as a launching pad for the individual who seeks detailed knowledge beyond what the book provides.

The ultimate objective is that the book will generate enough curiosity to stimulate further study and research into the issues raised in pursuit of seeking bridges between the indigenous modes and contemporary technological developments in the interest of effective communication.

Its theoretical focus is eclectic. The reader is thus exposed to more than one viewpoint on the issue. This is an opportunity for those seeking critical interpretations of social action as pertains to communication in a variety of African socio-cultural situations.

The book is the first comprehensive and collaborative attempt at a survey of indigenous modes of communication in Sub-Saharan Africa. The authors are African or non-African scholars with African experience. The content includes inventories of different modes of communication with theoretical discussions of their structures and potential. Empirical accounts demonstrate the possibilities and limits of the application of the indigenous communication systems to the design and implementation of education and development projects in the affected social systems.

K. Ansu-Kyeremeh
Accra, January 2004

ACKNOWLEDGEMENTS

In a collaborative work such as produced this book, it is the joint effort of the individuals involved which facilitates success. It takes a great deal of commitment to meet common deadlines, guidelines and thematic organization. That is why the greatest credit for the book goes to the authors of the various chapters. It is their cooperation, patience and ability to tolerate the editor's lapses which has finally found this book materializing.

Some of the chapters have been previously published in some journals. Wherever, this is the case, it has been indicated in a note at the end of the chapter. The original publishers who have given their permission for their articles to be republished as parts of the book are, however, acknowledged here.

All the chapters, even in cases where they have already been published as articles were edited to reflect the theme and structure of this book. One of the republished chapters, which appears as Chapter 9, "Training African Media Personnel: Some Psycho-Cultural Considerations" by Louise M. Bourgault however, is reprinted with gratitude to the editor and publishers of *Africana Journal* (1994, vol. 16, pp. 51–65). As with the other two republished chapters, the style in areas such as citations and references is reorganized to fit a common format.

The editor particularly acknowledges the cooperation of the *Journal of Development Communication* in allowing one of its articles, "From ritual to theater: Village based drama for development in Africa," authored by Louise Bourgault, which appeared in the Journal's 1991, volume 2, number 2, edition pp. 49–73, to be republished here as Chapter 6.

Chapter 8, "Indigenous Resources in a Ghanaian Town: Potential for Health Education," by Marie Riley is also republished with the permission of, and thanks to, the *Howard Journal of Communications* (Spring 1993) in which it originally appeared.

Chapter 2, "In search of a role for the African media in the democratic process" authored by Professor P. A. V. Ansah, was originally presented as a paper at the 6th Biennial Conference of the African Council on Communication Education. It is acknowledged here that the version published as part of the conference proceedings titled *Communication and the Democratic Process in Africa* is almost identical to the chapter. The main reason for this was that the author was unable to complete a major review he was undertaking. The Council's earlier publication is acknowledged with thanks.

The editor also acknowledges Mette Shayne, a Librarian with the

Africana Collection at the Northwestern University Library (Evanston, Illinois), for her special role as an electronic link between the editor and some of the contributors and also in her assistance through e-mail bibliographic resourcing. Mrs. Gertrude Missodey, Ghana Universities Press, Legon, Ghana, equally deserves special mention for so diligently and painstakingly typesetting the manuscript.

Although the different chapters are to be credited to their respective authors, the editor accepts overall responsibility for any errors in the final text.

INTRODUCTION

Kwasi Ansu-Kyeremeh

In the study and research of Africa, the "indigenous" dimensions of the social life of her peoples tend to be marginalized. By indigenous is meant the autochthonous social and cultural attributes and practices that characterized pre-colonial — and are still in many cases features of the social and cultural conditions of — peoples in societies across the continent. There is little recognition of the extent to which the forcible superimposition of Western socio-political structures and values during colonization has suppressed or obliterated valuable indigenous institutions and practices. The impact of westernization on the basics of life including food, shelter and clothing is phenomenal; and acculturation, assimilation and their resultant corrosion of local cultures which in turn have instigated cultural discontinuities and social dislocation have created confusion in education, politics, and every aspect of the African life.

In the field of communication in particular, and more so in mass communication, indigenous forms and dimensions of communication are often dismissed as "inconsequential," or only casually mentioned in the mainly eurocentric mainstream research. There are usually descriptive accounts of certain forms of "indigenous communication systems" without any theoretical and in-depth discussion and analysis of how those systems relate to the larger issues of social relations and interaction within the societies concerned.

An attempt is made in the next chapter to define the term "indigenous communication systems." It is to be understood from the outset, however, that the term is used interchangeably in this book with others including "indigenous media," "traditional media," "folk media" and "alternative media" to accommodate the diverse conceptual approaches of the various authors. In the particular case of "indigenous media," the context in which it is used here ought to be differentiated from the "native" context in which it is employed in places such as Canada, the United States, Australia and New Zealand by, for example, Ginsburg.[1] There, "indigenous media" usually refers to technology-based media as applied within, by or for the indigenous communities.

Regarding the tendency to marginalize indigenous communication systems in communication research in the African context which is of immediate concern here, one can assert that predisposition is no accident. Indeed, a number of reasons explain the limited scholarly attention paid to those systems. First, the discipline of communication study and research itself is so young that, by the time it gained legitimacy (as a disciplinary activity), the world was already in "the age of modernity" and was witnessing "the passing of traditional societies." With social research generally preoccupied with the fact of change and not necessarily the nature or content of change, communication research began with the perceived irrelevance of forms of communication that were viewed as "traditional." Consequently, it made sense for scholarship and intellectual activity pertaining to the "traditions" of communication in Africa to be affected by the eurocentricity (otherwise interpreted as modernity) which so inspired early scholarly investigations of African societies. As a matter of fact, the misrepresentation of indigenous communication systems as an anachronism is still maintained through the notion of "colonial mentality."[2]

The eurocentric claims that "traditional values and lifestyles are wrong, outmoded attitudes and systems which should be substituted by requisites of modernity,"[3] drove, and continues to drive, non-Western cultures, African cultures included, to engage in a race to achieve "modernity." An essential intellectual feature of this race is the desire of many indigenous African scholars to gear their research toward the justification of the need for Western technology-based media as necessitous innovations, or as substitutes for various African communication systems. At the centre of today's "post-modernist" world of computers, direct-satellite broadcasting, talk-back radio and interactive television is the "information superhighway" which depends on ever changing technology. The transformation of communication technology is a daily occurrence. In such an environment, not much of the research support outfit — international sponsor organizations, scholarly journals, publishers, or governments upon which researchers depend for financial resources and exposure — is interested in "traditional issues." This further makes the study of the field of indigenous communication unattractive.

Yet the limited available empirical and other evidence point to the existence of highly utilitarian and effective indigenous communication systems across the continent. Drama, storytelling, proverbs, poetry, and other such indigenous forms of communication are found to be good channels for communicating effectively, especially among the predominantly rural populations. For those peoples, the indigenous forms of communication are known to enjoy certain advantages over the technology-based media systems that may be introduced there. These advantages include structures that are fully integrated into the interactive holistic social systems of the societies concerned. This is unlike the technology-based media which are introduced extraneously and require the transformation of the people together with their lifestyle to suit the format and character of those media systems for any successful application. Furthermore, the indigenous systems can serve to meet the communication, social interaction, information, educational, development and entertainment needs of the people.

This book, which blends the theory of, and experiential accounts on, the subject is organized to deal with many of the aspects of indigenous communication systems in Africa. It explores theoretical perspectives against which the empirical evidence from some case studies that are reported can be examined to achieve a greater understanding of the issues. Time and again, the apparent reproduction of Western media forms and practices in prototypes all over the continent is criticized to expose their shortcomings while endeavouring to present alternatives in the local communication patterns. All the authors take an optimistic view of a crucial role for indigenous communication systems in development and education efforts across the continent; although in many cases the theoretical positions are not necessarily positivistic because some are essentially critical.

The book especially serves as a useful introductory material for students and instructors. The material presented also provides opportunities for the development of practical models by administrators and field officers of both local and international organizations that seek to understand the nature of unsophisticated and inexpensive communication in the African context. Furthermore, it is a resource in development communication and cultural studies, as well as reference tool for development educators and personnel of non-governmental organizations (NGOs) operating in Africa. In addition to their collective status as African or Africanist communication teachers and researchers, the contributors have multidisciplinary backgrounds that further enrich the text. However, the collective theoretical approach following from the multidisciplinarity is eclectic.

The chapters are organized into four parts. Part I focuses on the theoretical perspectives on indigenous communication systems. The authors in this part analytically examine the theoretical background to the structures and processes of indigenous African communication systems. Some explain, for example, how certain communication theories could promote or interfere with the application of certain systems in the design and implementation of development projects. Descriptive accounts and/or taxonomies are also presented in this part. Part II on the other hand explores the limits and possibilities of the applications of the indigenous media in development programmes and projects. Part III is composed mainly of specific case studies which demonstrate through empirical observations the processes of the indigenous communication systems which are discussed. Lastly, Part IV comprises of chapters which articulate future directions for upgrading the profile of indigenous communication systems as well as their incorporation into mainstream communication in Africa.

Ansu-Kyeremeh opens the discussions by defining "indigenous communication systems" in the African context. Also outlined in that opening chapter are the characteristics and possible applications of the indigenous systems as a communication genre. Three approaches to the application of indigenous communication systems in Africa by non-indigenous organizations and professionals are identified. These are the *sankofa* or retrieval approach (which is endogenous in every respect and considers indigenous systems as part of a holistic renaissance); the adaptation approach (which seeks a supplementation of the indigenous to the technological); and the synchronization or hybridization approach (which gives equal weight to the indigenous and the technological in various fusions). It is explained that the status of the indigenous systems, that is whether they are considered in policy as core elements in a general communication apparatus or as peripheral supplements or add-ons to the technology-based media, is largely determined by the prevailing theoretical standpoint and ideological leanings of those undertaking the desired interpretations. It is hypothesized that whether one adopts the functionalist viewpoint or the critical approach does not seem to affect the status of the indigenous systems because they are viewed from both theoretical positions as belonging to the periphery of communication issues. In other words, those systems are "traditional" and anachronistic from the perspective of both functionalist-diffusionist theories and critical theories.

Ansah underscores the need for further study of indigenous communication systems in the African context. His is a discussion of the perpetuation of the fallacy of an Africa without a tradition of free expression and human rights. He contends that development is achievable in a climate of freedom of expression. Ansah proceeds to identify various indigenous institutions, structures and processes that facilitated the freedom of expression in the powerful Akan Kingdom of Asante, which was subjugated by the British only after years of struggle between the two.

Ansah's chapter is followed by an overview, or a taxonomy, of a number of indigenous communication devices in Africa. Based on a survey of, and field interviews on, modes of communication in Nigeria, Wilson presents information on various forms of communication the characteristics of which are likened to what exists in other parts of the continent. Other groups of people on the continent, including Cameroons and the rest of the west coast as well as in east and southern Africa, he believes, share the Nigerian experiences. Wilson also classifies the various systems of communication into the instrumental media, the demonstrative media, the iconographic media, the visual media, the extra-mundane media and the institutional media. These classes are defined along with their forms and the purposes for which they might be utilized. Additionally identified in the chapter are the advantages of those systems when compared to the technology-based media.

At the beginning of Part II, an old experimental study conducted by Bame in which the relative responses to the technology-based media and indigenous communication systems by some villagers were measured is presented. The chapter was added because not many such studies are undertaken on the continent. The 1975 study is actually reviewed to take account of contemporary developments. One recalls how the mid-1970s was the period when critical theory was emerging to question the longstanding assumptions of the functionalist-diffusionist modernization theories which prescribed technology-based media as the only channels for "diffusing" development information in less industrialized societies of Africa. Bame's study was actually inspired by Rogers' diffusion model of the "dissemination" of development information.

Kees Epskamp profiles projects involving the application of local drama to stimulate development activities in Nigeria, Botswana, Malawi and Namibia. Based on these experiences, he makes a case for theatre for development and empowerment of development support communication in Africa. Through a descriptive historical analysis, he explores the application of theatre as an educational tool in the processes of social change. Emphasizing the importance of endogeneity in the design and process of development communication in the African context, Epskamp notes that some of the earliest post-war applications of theatre for awareness-raising were initiated and developed by expatriates even if the schemes were endogenous in the sense that the processes were intracontinentally generated.

Often, whether there are sufficient communication elements to separate ritual from theatre, and indeed, whether the two forms of expression are one and the same, is considered in discussions relating to the indigenous communication systems of Africa. Bourgault examines the use of village based-drama for development in mass communication in Africa in an attempt to resolve the question of whether ritual and theatre or drama can be clearly separated from each other in the African context. This is an important issue because its resolution could be helpful in determining the method and educational value of a communication activity as against the entertainment or religious significance of the same activity.

Morrison also discusses theatre. Her chapter looks at the theory, roots, and structure of participatory theatre in Burkina Faso. She notes that, as an interface between the village Burkinabe and the state, theatre plays an important role in social development processes.

Chapter 9, the second contribution by Bourgault, is a reprint of a recent journal article of hers. In it she expounds what she calls "psycho-cultural considerations" necessary for the training of African media personnel. She draws on her experiences in Nigeria and Tanzania to craft an analysis of African values and lifestyles and how they may promote or impede training programes for African media personnel. Prospective designers of media training programmes may find the discussion a valuable cultural analysis that gives insights on developing communication models to fit training programmes and processes.

Sometimes, an impression is created, by the concentration of the indigenous aspects of communication research in Africa on theatre and drama, that the two are synonymous with indigenous communication, or are the only identifiable examples of the genre. The definition of the term in the next chapter, as well as Wilson's taxonomy in Chapter 3, challenges any such impression. Still, at least three of the chapters — by Epskamp, Bourgault, and Morrison — focus on theatre or drama. Against the trend, Riley, who studied the indigenous resources of communication in a Ghanaian town, devotes her chapter to the identification of forms of indigenous communication other than drama or theatre. She includes an inventory of these systems, and proposes how they can be incorporated into community and social development.

Familiarity with power relations and how they relate to communication patterns within African political systems are as important to development communication planners as any other aspect of development activity. Identifying decision and opinion leaders is of little use if one cannot figure out the most effective ways of communicating with or through them. Part III discusses the political dimensions of the processes of certain indigenous forms of communication in some specific African communities. That part includes a chapter on political communication by Ansu-Kyeremeh. Discussion of two models of adaptive forms of indigenous communication in Uganda by Nabasuta Mugambi and in Burundi by Ngabirano complete the section.

Ansu-Kyeremeh's detailed description and analysis of the structure and processes of a Bono model of Akan political communication illustrates the contention of Ansah that there existed free flow of information in many pre-colonial African societies. Outlined in the Ansu-Kyeremeh discussion are the elements and principal actors in the local indigenous communication processes.

Nabasuta Mugambi articulates a feminist interpretation of the social and political significance of songs of some kind of operatic rendition as communicative tools in the transformation of the female status by Bugandan women's clubs. She constructs the songs and the mode of their performance as instruments for consciousness-raising among women in their pursuit of economic and social liberation from their men.

Addressing the question of form, Ngabirano, on her part, shows an example of the adaptation of local communication modes to the radio format to construct entertainment forms that promote "edutainment." To Pask (1976) education is entertainment; even if in the opinion of Pryor (1986) the assumption that educational television can work only by imitating the formats of entertainment television is a fallacy. The concern is not unrelated to separating the fact within an educational content from the fiction of an entertainment content which then becomes the issue of contention. The debate is akin to Bourgault's attempt to separate ritual from theatre.

It seems though that it is not so much as to whether entertainment has educational value as it is of the educational contribution a form of entertainment is capable of. After all, throughout the book, evidence is provided showing how various indigenous forms of entertainment are vehicles for educational communication. The question is whether whatever educational capabilities inherent in a form of entertainment restricts it to communicating only tasks related to social and moral learning and not to cognitive skill or motor skill development. Finding the answer to this question is, however, beyond the scope of this book; and so is a comprehensive examination of the significance of form and content of the communication system.

However, by focusing on a specific genre of communication, that is indigenous communication systems, the book, in its thrust, somehow acknowledges that "the medium is the message." The inculcation of European values was sought through European drama forms performed in European languages in mission schools as noted by Epskamp. Form, however, becomes less significant if missionaries can use local war dances in Podoland to teach christian religion. Ngabirano actually recounts an observation by one of her interviewees who thought an indigenously inspired soap opera ran "the risk of turning into a tool for manipulation and propaganda" by the ruling elite. Nonetheless, the intrinsic two-way communication formats of indigenous communication systems tend to elicit greater audience feedback and participation in the communication process compared to the largely one-way technologically mediated communication formats. That particular attribute is likely to minimize such manipulation.

Nabasuta Mugambi's discussion centres on the small-group Mityana women's club as a platform for transforming marginalized orality into song composition, choreography and even handicrafting as channels which communicate the messages of economic independence, social activism and self sufficiency to disempowered women and children in Uganda. Ngabirano, on the other hand, reports on preliminary findings of an ongoing case study of *Ni Nde?*, a radio soap opera in Burundi. The latter is perhaps the one study which exemplifies the attempts often made to seek African equivalencies to Western concepts. In this instance, a Burundian notion is presented as a variety of the broadcast programme genre, "soap opera." Tunstall (1977: 59) cites the Latin American *telenovelas* as an example of hybrid forms of communication, or "older cultural forms which continue in vigorous existence, although modified by Western influences." The problem with these models of communication, though, is "whether such hybrid forms are primarily traditional and 'authentic' or whether they are merely translations or imitations of Anglo-American forms" (Tunstall 1977: 59).

In later chapters, Awa and Wilson explore approaches to seeking a fusion of the indigenous communication systems and the technology-based media. Awa explores the foundations of the pessimism with which the indigenous communication systems are often perceived. The exploration is extended to an attempt to demonstrate in the chapter how even against the trend of the negative perception of the indigenous forms of communication and in the midst of the technological revolution, one ought to be able to find a role for the indigenous systems in contemporary Africa.

Concluding the chapters, Wilson outlines ways in which the indigenous communication systems may be considered by African nations in future communication policy formulation and strategizing. His discussion centres on the cyclical, inclusive and empowering nature of the communication process in traditional society which encourages feedback and audience input at all times.

This single volume has two new Chapters, 15 and 16.

In effect, the perspectives on indigenous communication systems in Africa articulated in this book serve three basic purposes. First, together, the discussions underscore the importance of indigenous sociocultural characteristics to all forms of communication processes that exist or are introduced into African social systems. As Bourgault explains in Chapter 9, knowledge of the culture of the African setting for which communication is being designed can be useful for the successful training of media personnel on the continent. Although the general theoretical approach may have positivist empiricist undertones, sometimes sounding paternalistically patronizing, the various literature reviews especially include critical analyses that provide an objective assessment of the issues discussed.

Secondly, a rich inventory of indigenous communication systems is provided as material with which one can work; or as pointers to what such resources may be available within societies which are not directly examined here.

Third are the diverse attributes of the indigenous communication systems which are outlined in different chapters. For example, beyond the cultural transmission capacity which is evident in Mugambi's discussion, the potential benefits of the systems as vehicles or channels for communication that facilitate the education of African in development propelling skills and attitudes in various aspects of life are apparent.

Indeed, communication is often linked to development in its capacity as a vehicle in the process of educating people about issues, concepts, techniques and skills that are necessary for pursuing actions or activities the outcome of which constitute some form of improvement in the lives of a given community. The subject of this book thus tends to directly or indirectly portray the role of communication in education geared toward the stimulation of development.

For example, Epskamp and Bame focus on the general development (actions and activities geared toward the improvement of lives of people and communities), potential of the indigenous communication systems. Riley, Morrison and Bourgault explore particular health issues including primary health care, family planning and AIDS. Bame and Epskamp respectively also address family planning and AIDS. Ngabirano's main concern is with education. Ansah and Ansu-Kyeremeh examine the political aspects of the indigenous communication systems. All the chapters incorporate case studies or experiential observations that serve to illustrate the character of the indigenous communication systems when applied to development programmes.

The book also includes a feminist perspective on indigenous communication. As indicated above, the chapter by Mugambi particularly focuses on feminist interpretation of a form of indigenous communication system. Other chapters such as the ones by Epskamp, Morrison and Riley discuss the interface between womanhood and various forms of indigenous communication systems.

With all this information, one hopes the book will provoke further discussion of the problems raised and help project the issues of the indigenous communication systems to the attention levels accorded the application of technologically mediated communication systems in Africa.

During the two years that the project to produce this book ran, the editor kept a close watch on developments in the area of attempts at an inclusive redefinition of the mainstream media in the African context that will incorporate the indigenous communication systems. Not much seemed to have happened. Actually, critical theorists such as Althusser and Gramsci, and even Paulo Freire (who prefers populist interpretations) all of whom tend to emphasize society, its structures and processes as a holistic entity, will hold that any such idea is wishful thinking. Their contention will be that the westernization (interpreted by some as modernization) that has engulfed Africa, is strongest in its adoption by a small section of the society. Into the hands of that few is concentrated all power and privilege. With this power, they seek to construct and maintain everything else that is Western, including the communication systems, through which they pursue the indoctrination of the underclass so that the former can maintain its grip on power and privilege. In the minds of the powerful then, the media will always be the technologically mediated channels of communication; and it is this definition that will always guide their planning of information dissemination.

The flipside of this thinking is to denigrate and show disdain for the indigenous social characteristics and traits, including the indigenous communication systems, which are then perceived as obstacles to modernization. This stymies research and articulation of the indigenous systems and by implication further reinforces the quest for modernity.

Bourgault's discussion in Chapter 9 inadvertently raises issues that echo the "anachronism" of indigenous sociocultural characteristics in the "modern" world. Bourgault, drawing attention to cultural and psychological factors trainers of African media practitioners need to be aware of, sometimes creates the impression that certain ingrained cultural and behaviour traits make it impossible for the African to develop the attitudes that are required to operate the Western technology-based media systems. One might ask: Why is it that technology is not designed to accommodate the cultural and social characteristics of African environs?

Other issues Bourgault raises in Chapter 9 may also be controversial.

Her discussion incorporates characterizations of the individuals in the study in a manner that some will consider as echoing the eurocentric modernization view of communication and social change propounded by Daniel Lerner, Alex Inkeles and Lucien Pye in which "traditional" societies were expected to "pass". In the extreme, some will even say it echoes the "noble savage" thesis of some Western scholars. Those who are familiar with the works of Andreas Fuglesang and others who see cultural factors not as antithetical to linear Western communication thought systems as propounded by Lerner and the others but as alternatives in a multidimensional thought system in a complex world are likely to criticize and reject, rather than accept or adopt the explanations and positions postulated by Bourgault.

Whichever way one looks at it, from whichever ideological or theoretical perspective, "interactive" has become the "in-concept" in communication in the 1990s. Those who have cared to examine the indigenous communication systems in Africa have all come to admire and appreciate the intrinsic "interactive" quality of those systems. This suggests that, in the pursuit of "interactiveness" in the realm of technologically mediated communication, it might be useful to re-examine the principles which guide the indigenous communication systems so as to determine whether those principles provide clues for the 1990s phenomenon.

One may need to state that the investigation into the structures, capabilities and processes of the indigenous communication systems in Africa is significant not just for a historical record of descriptions and inventories. It might be important to know that there once existed, or still exists, a drum for conveying certain messages. To develop a contemporary relevance in the context of technologically mediated communication, however, one needs an understanding of the means of ownership of the drum, the mechanics of its message formulation, its message "dissemination" or sharing processes as well as access to both its message formulation forms and the content of the message. This will help extend any qualities of an indigenous system that could have some relevance for contemporary communication to be utilized as such.

Ansu-Kyeremeh (1994) has articulated an *indigenization* of communication in the African context, a formulation that transcends cultural symbolism. Together with the issues raised in this book, they should provide a basis for further investigation into the contemporary as well as future significance of Africa's indigenous communication systems in supporting education and development on the continent.

In Chapter I, it was stated that without an intention to open the debate over the autonomy of communication systems which borders on the effects issue in communication, one would like the indigenous communication systems to be considered more in terms of application and utilization rather than their ability to perform functions. This is because, as Epskamp noted, a specific indigenous communication system can be manipulated to serve different purposes by the source from which a message channelled through it originates. Bame also believes that "a real danger" exists for folk drama to be "co-opted by interest groups such as the ruling class in a country to foster their own interests, for example, for maintaining themselves in power." Such a conceptualization of the indigenous systems differs from general media effects theory in that, because of the instant response capability, the built-in feedback attributes of some of the indigenous systems are such that the response from the receiver can as much influence the ultimate meaning formulated into a message as the sender. Bame actually advices popular theatre experts to work with the people at the grassroots level if they are to counter the co-optation of theatre to serve ruling interests.

All told, the challenge posed by this book is whether communication research in Africa will begin to take indigenous communication systems seriously and thus intensify attempts to investigate them to establish the programme of their relevance to contemporary development planning and programme implementation.

Thus, in no way is this attempt to highlight the importance of the indigenous communication systems to be construed as representing the systems as panacea for Africa's development communication problems. That will be far from the objective of putting the book together.

NOTES

1. Faye Ginsburg, Indigenous media: Faustian contract or global village? *Cultural Anthropology*, 6(1), February 1991: 93–112.

2. The expression "colonial mentality" is employed to describe the tendency to conceptualize or act European, such as adopting a cultural artifact, idea, behaviour, or attitude where a clear indigenous African alternative exists; it is an indicator of dependency. A reader's letter to *West Africa* (24–27 Dec. 1992, p.2177) noted "inviting foreigners . . . to oversee electoral proceedings . . . smacks of the old colonial mentality." D. Mbachu (*West Africa*, 1988, p.2135) cites DJs playing too much music from the West as another example of colonial mentality.

3. Andreas Fuglesang, "The myth of a people's ignorance," *Development Dialogue*, 1–2, 1984, p.47.

PART I

THEORETICAL ISSUES

Chapter 1

INDIGENOUS COMMUNICATION IN AFRICA: A CONCEPTUAL FRAMEWORK

Kwasi Ansu-Kyeremeh

Introduction

When critical theory emerged in the 1970s to challenge the assumptions of the then widely-accepted functionalist explanations for development communication, part of the effort was directed toward a search for alternative communication systems. The search included the examination of indigenous communication systems, sometimes through experiments, such as the one described in the chapter by Bame, in which those systems were matched against technological media to determine the attention paid to each group and their effectiveness as channels. While such efforts have endured into the 1990s in Asia, in Africa they have hardly been sustained. Yet the continent finds itself increasingly in need of alternatives to technology-oriented communication systems which continue in their rapid innovation to outpace the continent's technological capacity.

The paltry literature regarding various interpretations of indigenous communication systems elsewhere and in Africa is briefly reviewed in this chapter. The review is organized into an initial definition of indigenous communication systems, an examination of their characteristics and attributes as well as purposes for which they could be made to serve society. Three approaches to the conceptualization of the systems are also explored. For now, there is a brief look at the literature including an overview of the various conceptualizations.

Definition of the Indigenous Media

One of the purposes of this discussion is to open up the narrow definition of "media" beyond the technology-oriented communication systems to include endogenous communication channels. First, the term "indigenous" is interpreted here as "that originating from a specific place or culture."[1] This implies that anything indigenous needs to be qualified in terms of its place or culture of origin.

Scholars who employ the term "indigenous" to distinguish endogenous media systems of non-Western cultures from Western mass, or modern, media include Wang and Dissanayake (1984), Jussawala and Hughes (1984), and Ugboajah (1985c). Writing from the perspective of Asia, the social conditions of which are in many ways similar to those of Africa, Wang and Dissanayake (1984: 22, 27) preferred the term "indigenous communication systems" to "indigenous media," so as to avoid what they perceive as "confused" usage of the terms "folk media" and "traditional media." Recalling a Ranganath (1979) categorization of the folk media into ritual, historical and traditional, as well as utilization, they believe "folk media should apply mainly to the performing arts" which include "puppetry, shadow plays, folk drama, folk dance, ballads, and storytelling." On the other hand, "traditional media," in the view of Wang and Dissanayake (1984) are "interpersonal channels and networks of communication, such as the Indonesian *banjar*, the Korean mother's club, and the Chinese *hui* (loaning club)." The term indigenous communication systems thus encapsulates folk media and traditional media. Wang and Dissanayake (1984) further describe the indigenous communication system as "ingrained in the culture" of its host community.

Jussawalla and Hughes (1984) also employed the term indigenous communication systems to denote channels that are "embedded within [the] traditional mores of a people and contributing significantly to their history and culture." Without adopting the Wang and Dissanayake (1984) separation of the folk from the traditional, Jussawalla and Hughes (1984: 255–6) define the indigenous communication systems as

> . . . those systems of communication which have relied historically on informal channels to convey information and which obtain their authority from the cultural mores, traditions, and customs of the [people] they serve.

From an African perspective, Ugboajah (1985c), who also uses the term "man-media" (1986: 17), introduces another term "oramedia,"[2] In addition to the two terms, he also uses "folk media," "traditional media," "informal media," and "indigenous media" interchangeably. It is important to note his definition of "oramedia" as "grounded on indigenous culture produced and consumed by members of a group." Also in the African context, Hachten (1971: 171) referred to the indigenous types of communication as "informal channels of communication."

Conscious of all these interpretations of the indigenous communication systems, they are conceptualized here, inclusively, as:

> Any form of endogenous communication system, which by virtue of its origin from, and integration into a specific culture, serves as a channel for messages in a way

and manner that requires the utilization of the values, symbols, institutions, and ethos of the host culture through its unique qualities and attributes.

In this definition it is assumed that the indigenous communication systems are incorporated into a specific culture and that they have a better chance of effectiveness within that culture than other exogenously-originating media systems. Indigenous media and indigenous communication systems are used here interchangeably. A third one, "communal media" is also used at times.

The meaning of communal media, akin to what Stover (1984: 68) calls "community media," is widened to incorporate externally originating media systems which demonstrate congruency with the structures of the local culture in such a way that they are amenable to an adaptation to the social organizational structures of the host culture for maximum optimization of use. Communal also connotes Gyekye's (1987) explanation of a combination of individualist and group predispositions but not in the sense of "groupthink," or homogeneity in thought, as implied by Janis (1972) and Pratt and Manheim (1988). Yet another term, "local media," is also used, although sparingly for convenience, in place of indigenous media.

In his thesis, Ansu-Kyeremeh (1989) grouped the indigenous communication systems in a rural African context into venue-oriented communication, events as communication modes, games as communication channels, and performance-oriented communication and he later provides examples in this book in his discussion of modes of Akan political communication. These may differ from Wilson's classification.

However, whatever the classification and irrespective of the terminology, descriptors, or interpretations, a number of uses to which the indigenous media could be put have been identified. These will be examined now.

Indigenous Media in the Developing Context

Fuglesang (1979, 1984) and others have devoted much of their research efforts to identifying the significance of natural communication modes among members of rural communities. Such research includes the one by Wang and Dissanayake (1984) in Asia. Parmar (1975) and Malik (1982) surveyed the situation in India where a comparative analysis of the performing arts in education for development was recently undertaken by Kidd (1984b). Also in the Asia-Pacific region, the subject has further been explored by Abrams (1984) from the Papua New Guinea perspective, and Valbuena (1986) on the Philippines.

In Africa, Doob (1961) surveyed "traditional" forms of communication even before the critical 1970s. Later, Opubor (1975) included items on the theatrical forms of communication in his bibliography. Kidd (1984b) highlighted popular theatre by focusing on the folk theatre *Laedza Batanani* as an experiment in nonformal education in Botswana. Epskamp later gives more detail about the experiment. Also in the eighties, Ugboajah (1985, 1985c) undertook one of the few surveys of the structures, characteristics and uses of the indigenous media in Africa. Storytelling for education and development is sometimes the focus of communication research as was the case with Fiofori (1975). More recently, Riley (1990) explored the untapped potential of indigenous resources for promoting primary health care with particular emphasis on West Africa. Her chapter, which appears later, includes an inventory of a variety of indigenous communication resources.

A study by Bame in which he investigated the role of indigenous comic plays in social change is reproduced in this book. Other works such as those of Opubor (1975), Ugboajah (1985), as well as the main thrust of Pickering's (1957 a&b) study, all explored the role performed by the various local forms of communication in the village situation.

The literature on indigenous forms of communication shrinks though when it comes to the generic treatment of indigenous communication systems. Only Doob's (1961) research seems to have attempted to cover cross-continentally the various forms of the indigenous media. With so much emphasis on the performing arts or drama or folk theatre, including the chapters by Bame, Bourgault (Chapter 6), Epskamp and Morrison, as observed by Riley, an impression is created that only those forms of communication constitute indigenous communication systems.

The few works encountered by Ansu-Kyeremeh (1989) that identify the indigenous communication systems in Africa included analytical empirical studies, mainly of the summative evaluation and project report type, and were basically in the form of experimental or pilot micro studies. These included works by Gill (1984) and Smart (1978). Some, such as the works of Ansah (1985) and Ripley (1978), were of the essay or lecture type. There were also the descriptive experiential accounts such as those by Boafo (1984) and Obeng-Quaidoo (1987).

Another common feature of most of the empirical research, evaluation reports, experiential observations by practitioners, and critical essays is the emphasis researchers placed on the inherent deficiencies of the Western mass media when applied to the African setting. In other words, they do not probe the strengths of the indigenous communication systems. This is in spite of the widespread belief that an inventory of all available media and communication systems, with their identified attributes, is necessary for judicious, relevant and rational media use in decision-making (Reiser and Gagne 1982). Empirical studies also hardly explore the enabling characteristics of indigenous communication systems.

Ripley (1978), as does Bame and Riley, however, demonstrated greater sensitivity to the cultural environment. Bame's (1975a; 1975b) study provided empirical evidence to support the assumptions Ripley (1978) put forward in his lecture. In that study, Bame, as he reiterates in Chapter 4, discovered that family planning educational messages presented by the indigenous modes of communication, such as comic plays or *concert* and group discussions, were more favourably received by villagers than those presented through the technological media of radio, television, and the cinema.

Next, the characteristics of the indigenous media that make them so enabling in village conditions are outlined.

Characteristics and Attributes

It is generally believed that the characteristics of the audience and the community, or the sociocultural system in which a medium operates, are paramount in every communication process. In Parmar's (1975) view, communication does not occur in a cultural vacuum; it is a derivative from the sociocultural context in which it takes place (Jefkins and Ugboajah 1986). Since cultures differ in the ways in which social relations are conducted — in terms of communication symbols and patterns — one expects different outcomes from the impact of different media with different characteristics. In such circumstances, audience characteristics, including their socio-economic and cultural milieu, may prove crucial in choosing the appropriate media for communication in especially the rural African environment. For these reasons, the attributes and characteristics of communication systems are examined in the context of the social interactive patterns, including their group and individual dimensions.[3]

Perhaps the most important characteristic of a communication setting is the facilitation of corrective feedback. For example, where the media are used to communicate knowledge Reiser and Gagne (1982: 505) are of the view that "corrective feedback is best provided by interactive media" which make possible differential feedback displaying the correct response as well as the degree of correctness of audience response. Trevino *et al.* (1987) and Markus (1987) share this view. It is in this context that many acknowledge that the characteristics of the target audience assume even greater importance in choosing appropriate media. When the familiarity and popularity of indigenous communication systems with villagers are considered, it follows that they will be legitimate factors in media choice. Indeed, in development education, it will be necessary for educators to be conscious that media may be differentially effective for different types of learners, and, therefore, media that are best suited for various learner types should be identified (Reiser and Gagne 1982: 505).

To be even more successful in choosing appropriate forms of communication, consideration will also have to be given to the characteristics of indigenous communication systems which enable them to be utilized in the ways identified above. That is why such characteristics, particularly the educationally enabling attributes, as identified and described in the literature, are briefly outlined in the next few paragraphs.

Generally, the enabling or disenabling characteristics of indigenous communication systems, which could enhance or diminish their communicative capacities, may be inferred from the works cited above. Examples are that they tend to be small group-oriented thus suiting the small rural audiences; are rooted in the local culture; are largely orally-based; and that they elicit two-way dialogical communication processes. They are also universally and cheaply accessible and they pose no (common) language problems. Also, the indigenous media are usually relevant in their structure and content to the host audience; they are interpersonal; and they have the capacity for instantaneous programme production and presentation (Ugboajah 1985c:167). Indigenous media are thus basically interactive, utilizing concurrently the senses of sight and sound in face-to-face encounters; and, therefore, maximizing active participation of audience in the communication process.[4]

In almost all the chapters in this book, favourable characteristics of the indigenous media which facilitate their applications are identified. Bame, for example, states as part of his findings that the "two folk media — concert-party plays — and village group discussions" he used in the study effectively combined visual and oral effects. The plays were familiar to, and popular among, the rural illiterate target audience; and their mode of communication could be deliberately transformed to be utilized in education, information, entertainment, and development. In his study, Ansu-Kyeremeh (1989) observed that because of their interactive two-way communication character, the village discussion groups allowed for a high degree of personal involvement by participants. The inclusive process created chances for the audience to participate actively to the point that it enhanced their commitment to the ideas and decisions reached during discussions.

Indeed, it is often felt that one of the most important features of a rural basic education programme must be that "learning would be chiefly an oral process, parallel to ways that social transactions take place." Such a process was actually observed in a Ghanaian situation by Ansu-Kyeremeh (1992). Manuwuike (1978: 18–19) also noted years ago that rural African communities in general are "orally articulative" in their communication processes and added that "the written word becomes the least effective means of disseminating information." The intrinsic characteristics which were just outlined enable the indigenous communication systems to be used in a number of educational and development situations. Some of the purposes for which they could be made to serve are examined next.

Purposes Served by the Indigenous Media

Ugboajah (1985c: 167) believed that the indigenous or "oramedia" "may be defined as functional and utilitarian," and that:

> Their most important purpose is to provide teaching and initiation, with the object of imparting traditional aesthetic, historical, technical, social, ethical and religious values. They provide a legal code of sorts which rests on stories and proverbs generated through the spoken word. They also play other roles in the village society such as mobilizing people's awareness of their own history, magnifying past events and evoking deeds of illustrious ancestors. Thus they tend to unite a people and give them cohesion by way of ideas and emotions.

Ugboajah (1985c: 11) further asserted that the "'oramedia' are great legitimizers because they are highly distinctive and credible."

Summarizing the utilization patterns of the indigenous communication systems in the Asian context, Wang and Dissanayake (1984: 27) noted that:

> The studies conducted so far on indigenous communication systems can be divided into three basic categories: indigenous communication channels as used in development programmes, folk media as forms of entertainment, and the cultural characteristics of folk media.

Post-exposure data gathered during Bame's (1975a) study of local comic or concert-party plays and village discussion groups pointed to the educational significance of these indigenous media forms. Rural respondents ranked the concert-party play first, from a group of both

Western mass and indigenous media, as the most important media source for their information about family planning. Another empirical study conducted by Fiofori (1975) found that "communoraldiction," popular tales, or oral narratives, all of the indigenous communication genre, dealt with situations familiar to listeners through an economically inexpensive method for spreading information. Fiofori (1975) concluded that the indigenous media had the capability to short circuit the time span that could be a hindrance for the people who needed the information; and provided a more convenient way of communication "as it is indigenous to traditional people and is also an integral part of traditional life." Bame also found a relationship between living in a rural community and relying more on traditional (indigenous) media as the source of family planning information. One wonders whether this linkage between the indigenous media and rural culture is responsible for the lack of mainstream interest in that area of communication research since the indigenous systems and rurality are both marginalized in national development efforts.

From India, Parmar (1975: 69) observed that techniques of storytelling such as "twisting" (that is, "incorporating in its [story] body a suitable message to keep pace with what the situation demands") facilitates adaptation to suit local needs. For Africa, though, Ofori-Ansa (1983: 5) recommended that

> . . . greater use should be made of indigenous channels of social interaction and consensus generation . . . to minimize the negative side effects of extensive local participation in [educational] project formulation and implementation.

These statements underscore the capabilities of the indigenous communication systems. That is not to say that the use of these systems is without limitations.

One needs to understand also that what are represented as the functions of the indigenous systems are in actual fact purposes for which they can be employed to serve. Without necessarily visiting the contention over the autonomy of communication systems, it is clear that the initiator of communication is at liberty to choose the channel through which his or her message is to be directed.

Limits of Indigenous Communication

Although indigenous communication systems possess qualities that make them suitable for use as means for communicating educational information, Parmar (1975: 70) cautioned that as far as storytelling is concerned, "tales must be cleansed of age-old errors, of religious mysticism, superstition, unseemly incidents and episodes, of false constructions." Ugboajah (1985) made similar observations. Doob (1961: 190) on his part found in a survey of communication systems existing in various African societies that some local customs got in the way of successful communication. Even then, the indigenous media, either individually, or as a communication network system, cannot be credited with the degree of autonomy assigned to the Western mass media by Von Hentig (1986) and McLuhan (1964). Media institutions in the West are perceived by the two scholars as assuming their own character and enough autonomy to influence action and thought of society. Indigenous media on the contrary are inextricably integrated into the whole social system. This quality intrinsically endows them with the capacity to respond to the local cultural milieu. At any rate, if the indigenous systems have any functional capabilities then they ought to be relevant to the social systems of which they are part.

Sociocultural Relevance

The belief held by Parmar (1975), Sonaike (1987), and Jefkins and Ugboajah (1986) that communication is inextricably linked to the socio-cultural institutions and human relationship patterns in which it occurs was referred to earlier in this discussion. Both White (1976) and Hall (1977) shared this belief. The cultural significance of the possible impact of technology-based media systems is further stressed by Stewart (1985), Ely (1983), and Aranha (1983). A common element in the thinking of these implicit "believers" in the need to pay attention to the indigenous communication systems is that they seem to agree that the definition of "communication media" ought to extend beyond the Western mass media types to include the indigenous communication systems. Their concern with the cultural contexts of communication seems not unrelated to the need to consider both technology-oriented mass media systems and indigenous communication systems in choice situations. The tendency to employ languages other than that which are locally known and understood (Ansah 1986; Stover 1984) is often closely associated with reliance on the technological media. Thus, in a situations where the host culture is such that the use of local languages predominate one would expect greater application of the indigenous communication systems. The significance of language is also underscored by Reiser and Gagne (1982) and others who believe that it is critical because not only is language very closely tied up with one's thinking about intellectual and social processes, but also the individual's "real world" is to a large extent unconsciously built upon the language habits of a group.

The nature and use patterns of the indigenous communication systems outlined above tend to influence perceptions of what the nature of the relationship between those systems and the technological media should be in Africa. Three main approaches to the construction of the relationship are observed by Ansu-Kyeremeh (1994). They are the *sankofa* approach, the adaptation approach, and the hybridization or synchronization approach. *Sankofa* regards communication as part of a necessary renaissance of African social value systems and cultural structures. Adaptation echoes dependency in that it relies on the technology-based media as the principal channels to which the indigenous systems must be adapted. Synchronization or hybridization considers the technology-based media and the indigenous communication systems to be of equal importance and therefore recommends the fusion of their positive and compatible attributes.

Summary

An overview of the literature on communication forms that are rooted in the social fabrics of indigenous communities in various parts of Africa was provided above. The discussion and analysis pointed to the different interpretations and conceptual approaches researchers adopt toward those forms of communication. The review indicates that the communication systems identified have certain physical attributes and intrinsic qualities which enable them to perform a variety of communication tasks within the host communities. It was, however, also evident from the literature that indigenous communication is usually analyzed from the Western perspective, thereby reducing the significance of the socio-cultural context of their existence.

Meanwhile, enough evidence exists in the literature on the potential of the indigenous media for education and development in rural communities in Africa. Even so, there seems to be no large-scale empirical or theoretical studies on their ramifications for the communities in which they exist. As a result, minimal efforts have gone into understanding the capacity of indigenous communication systems for various applications. This means that there is still the need to investigate them comprehensively, especially when one considers the observation by Bielenstein (1978: 16) years ago that "some of the traditional media could furnish a very significant contribution to the [ongoing debate over the] New World Information Order."

NOTES

1. The first of the two meanings assigned by the *Webster Comprehensive Dictionary*, Encyclopedic edition, Vol.1, (Chicago: J. G. Ferguson, 1977, p.644)

2. Derived from the fact that rural African cultures are orally based communities as noted by E. Manuwuike, *Dysfunctionalism in Afrikan education* (New York, Vantage, 1978, pp. 18–19).

3. Reiser and Gagne (1982: 504) observe that only five out of ten models included in their survey of media selection processes "require direct consideration" of factors such as the learning setting, learner characteristics, and learning task characteristics as important in determining what media is to be adopted in a learning situation.

4. Participation is vital in the communication process, more so   in media application to education. Indeed, this quality of the indigenous media can be the envy of the Western mass media which rigorously seek audience input through "letter to the editor" columns in the print media, and formats such as establishing access phone lines for "talk-back," "phone-in," "back-chat," and "forums" by the broadcast media to elicit audience participation. Even the cinema plays on audience relevance by trying as much as possible to get audience to relate to films; and the concept of teleconferencing" epitomizes the need for small-group oriented media systems in pursuit of the elusive participation.

Chapter 2

IN SEARCH OF A ROLE FOR THE AFRICAN MEDIA IN THE DEMOCRATIC PROCESS[1]

Paul A. V. Ansah

Even though "modern" mass media systems are not indigenous to Africa, there were certain accepted practices governing communication, and the expression of ideas in general, in traditional societies which can usefully inform the search for a role for the modern media. Ansu-Kyeremeh does provide an example in another chapter in this book. It thus need not happen, as Bourgault suggests in Chapter 9, that "Western normative values, including free press and objectivity . . . clash with local political realities or the local operating definition of 'development journalism.'" Such a "clash" may result only from the lack of interest in the two-way format of indigenous communication systems and preference for the one-way format of the Western technological media. In order, therefore, to determine the contribution that communication can make to the democratic process, it may be instructive to briefly find out what useful lessons may be learnt from African political and cultural traditions regarding human rights in general and freedom of expression in particular. We consider this background essential to the discussion because it will provide a response to those who argue that the concept of press freedom is alien to Africa and, therefore, a luxury that cannot be afforded at this stage of national development.

Human Rights in African Cultural Traditions

From the way in which human rights have been disregarded in Africa, the impression can easily be created that there is no tradition of human rights in Africa. There is, however, considerable evidence to the contrary. Despite the wide variations in the political systems of pre-colonial Africa, it is possible to make some safe generalizations on the basic principles of human rights. Whatever the different articulations and expressions of human rights, one common point in all the systems, such as is explained later by Ansu-Kyeremeh, was that they were infused with checks and balances. Even the king or chief was subject to laws, and failure to abide by them could entail his destoolment or removal from office. There was thus little room for despotic or absolute power over the citizens. There was a distribution of power, and the principle of prior consultations with councillors or elected representatives of the people was recognized. There was also a clear conception of the right to free self-expression and the freedom of association as attested to by numerous proverbs, court practices and oral traditions. The structure of traditional society itself was meant to guarantee civil and political rights. There was devolution and decentralization of authority such that, in practical terms, the king or chief was basically a *primus inter pares*.

The argument about the extent to which human rights formed part of traditional African political culture stems from the fact that with modernization, there is a general decline in the traditional constraints upon both the rulers and the ruled. Societies have become larger, kinship ties are no longer as solid and as binding as before, hence the need to formally define human rights and formalize the means for their enjoyment within the framework of the modern state.

It must be noted that the expression and formalization of human rights in any society will be governed by the level of modernization and instruments that the society has at its disposal. For example, the freedom of expression can be concretized in the freedom of the press in a society which has the facilities for printing. But the basic principle remains the same. Since the traditional structures and arrangements for ensuring a consensual use of political power do not appear to be adequate in the present circumstances, new mechanisms, structures and institutions have to be devised to ensure the promotion and protection of civil, political and personal security rights. It is this kind of concern which has eventually given rise to a document like the African Charter on Human and Peoples' Rights (OAU 1981).

It can be said, then, that if "modern" Africa's record in terms of the respect for human rights is a dismal one, this cannot be properly attributed to the assumption that Africans have a completely different conception of human rights from that of the Western societies with which they came into contact through colonialism. The reason has to be sought elsewhere, but certainly not in African cultural and political traditions. In traditional African society, freedom of expression was recognized as a fundamental human right where consensus was given a high premium, and this was based on the free expression of opinion. Citing an example from the Akan of Ghana, Busia (1967) writes:

> The members of a traditional council allowed discussion, and free and frank expression of opinions, and if there was disagreement, they spent hours, even days if necessary, to argue and exchange ideas till they reached unanimity.

This means that if freedom of expression has suffered diminution in Africa, this is due more to intolerance, the establishment of one-party or non-party states and military regimes ruling by decrees and edicts rather than to any traits inherent in African traditions.

In support of the contention that despite differences in the mode of expression human rights are seen to be truly universal and that they are also embedded in African traditions, one can do no better than recall that the African Charter on Human and Peoples" Rights is quite explicit in its preamble on certain points which it sees as deriving from African civilization and culture. With particular reference to the freedom of expression, the Charter states in Article 9:

1. Every individual shall have the right to receive information;
2. Every individual shall have the right to express and disseminate opinions within the law.

Chapter II of the Charter prescribes certain duties. Among these duties, sub-Paragraph 7 of Article 29 enjoins each individual

> to preserve and strengthen positive cultural African values in his relations with other members of the society, *in the spirit of tolerance, dialogue and consultation* and, in general, to contribute to the promotion of the moral well being of society (emphasis added).

The import of this provision is that governments are expected to establish institutions and create the atmosphere that will make it possible for citizens to contribute to the general development of the society, and this means guaranteeing and respecting their freedom of expression and of association, among others.

Press Freedom, Democracy and Development

It has been argued in some quarters that in the face of the enormous problems facing many African and other developing countries, it is necessary to restrict civil and political rights in order to accelerate economic development. The implication is that the promotion of human rights should be subordinated to the imperatives of economic development because the two objectives cannot be pursued simultaneously. This view appears to be based on an outmoded and mechanistic view of development which does not take the human factor into consideration. An enlightened view of development defines it as:

a widely participatory process of social change in a society, intended to bring about both social and material advancement including greater equality, freedom and other valued qualities for the majority of the people through their gaining greater control over their environment (Rogers 1976).

This enlightened concept of development recognizes the need for material advancement, but it also puts the emphasis on human dignity and the active involvement and participation of the people in the development process.

It is the outmoded, materialistic concept of development which does not reckon with the promotion of human rights as an integral part of human development. The new concept, on the other hand, implies that economic and social rights can, and should be, pursued concurrently with political and civil rights.

The new concept of development puts a premium on participation in the discussion of the affairs of the state as well as in decision making. The question then is: How can this participation be ensured in the absence of the right to express oneself freely and frankly? For the purpose of national development or self-development, people should be able to share ideas and discuss freely, exchange views, evaluate alternatives and criticize where necessary. One of the functions of communication is to provide an avenue for social interaction and participation. The mass media system of any country wishing to develop, therefore, should provide a forum or a platform for collective discussion and the weighing of various options in order to arrive at well-considered decisions. In other words, to serve the ends of development, the mass media should provide a market place for the exchange of comment and criticism regarding public affairs.

Authoritarian controls which preclude the free exchange of ideas have been imposed by governments of many developing countries in the name of the need for rapid development and this situation has led to a refinement of the authoritarian theory of the press into the developmental theory of the press. Though it is not easy to give a neat definition of this theory, one can isolate certain important characteristics relevant to the present discussion. The theory enjoins the media to carry out certain development tasks as defined by national policy makers; it is also based on the national policy makers, as well as on the notions that the need for collective development should take precedence over individual rights and freedoms. Under this theory, the right is conceded to the state to restrict media operations and to exercise direct control to ensure that media resources are used to promote national identity and integration. The authoritarian streak in the developmental theory is evident, and Lent (1977: 18) tried to rationalize the argument of some Third World leaders and scholars when he wrote:

> Because Third World nations are newly emergent, they need time to develop their institutions. During this initial period of growth, stability and unity must be sought; criticism must be minimized and the public faith in government institutions and policies must be encouraged. Media must cooperate, according to this guided press concept, by stressing positive, development-inspired news, by ignoring negative societal or oppositionist characteristics and by supporting governmental ideologies and plans.

What this amounts to is an enforced or imposed consensus in the name of national development, and this is clearly at variance with the more enlightened view of development which sees free participation as an essential ingredient.

But developmental theory of the press has also been seen within the context of participation.Ogan (1982: 11) considers that it can be interpreted to mean "the critical examination, evaluation and report of the relevance, enactment and impact of development." In order to undertake this critical evaluation, it is essential that the media be sufficiently free and independent of government control. This way of defining the developmental theory of the press brings it nearer to the libertarian and social responsibility theories than to authoritarianism. Echoing the same kind of sentiment, Aggarwala (1979: 181) states that the role of the journalist inspired by the development theory is to

> critically examine, evaluate and report the relevance of a developmental project to national and local needs, the difference between a planned scheme and its actual implementation, and the differences between its impact on people as claimed by government officials and as it actually is.

It is clear from the opposing definitions of the developmental theory that it is still unsettled and its real interpretation is still under debate. The difference in interpretation seems to arise out of whether one perceives it as a normative theory depicting how the media serve society, or as an

> objective theory which attempts to describe what the media actually do. It is from the normative perspective that Aggarwala (1979) arrives at his definition of the developmental theory, which seems to be inspired by participative communication models favouring democratic, grassroots involvement.

Closely echoing a liberal and democratic interpretation of the developmental theory, Anim (1976) puts forward the idea of a "participant press" system which he considers appropriate for the needs of developing countries. This participatory theory of the press, which is to be distinguished from the "democratic-participant media theory" described by McQuail (1987), is not seen within the context of either an adversary or servile relationship between the government and the press as Merrill and Lowenstein (1979) see it. Anim (1976) postulates that in order to serve the ends of development, the relationship between the government and the press can be characterized by cooperation and understanding without confrontation or servility. He puts forward ten principles underlying his "participatory theory of the press" among which, for the purpose of the present discussion, one can note the following:

1. A participant press would operate on a dialogical principle, providing the community with a platform for free and active discussion among its members and between them and the political leadership.

2. A participant press would conceptualize the process of national development as a search by all members of the community for viable solutions to problems which affect leaders and followers alike.

3. A participant press would operate on the principle that every member of the community is a searcher after the truth and that, until a consensus is reached, one member's decisions are as important as those of any member.(Anim 1976: 126).

It is easy to see that Anim's (1976) participatory theory is basically inspired by libertarian notions and recognizes the need for the expression of diverse opinions within the media system. According to his theory, therefore, democracy and development are compatible and not antithetical, even at the present state of development of African countries. He stresses that the role of the press should be one of education and information and not prescription, and adds that "social criticism and conflict are not necessarily disruptive. They are important factors in creating and maintaining a participant society and in contributing to sound national development" (Anim 1976: 133). Nyerere (1973) is obviously referring to the underlying principle of participative communication for development when he says:

Development brings freedom, provided it is development of people. But people cannot be developed; they can only develop themselves. A man develops himself by joining in free discussion of a new venture, and participating in the subsequent decisions; he is not being developed if he is herded like an animal into the new venture.

Government-Press Relations

The relationships between the government and the press in Africa have generally been characterized by tension and conflict. Where there appears to be no tension, the reason is that the press has either been cowed into submission or it has become an organ of the ruling party. What one sees is a shrinking of the privately-owned press and an expansion of the state-owned press.

This situation appears to reflect the intolerance and the tendency towards the creation of monolithic political institutions in Africa. There has been a systematic suppression of all organized opposition, and the elimination of all forms of organized dissent has usually paved the way for the establishment of one-party states or of military regimes. Just as the creation of one-party states has been justified on the grounds that it is necessary to mobilize all human resources within the nation for national integration and development, the virtual monopolization of the mass media has been explained in terms of the need to ensure that the people are not distracted by "false propaganda," and that all media resources will be harnessed and directed towards national development.

While people do not seriously question the government's ownership and control of broadcasting, they tend to see in government's monopolization or domination of the press a streak of authoritarianism and interference with the people's right to receive information from diverse sources. There is, however, another way of looking at the question of providing a justification for a government's ownership of a newspaper. The government needs to inform people about its plans and programmes and to mobilize them for development, using all the channels available. Whether elected or self-imposed, a *de facto* government, acting in the name of, and on behalf of, the people would seem to have a stronger claim to inform and educate them as well as to try to mould their thinking than the owner of a private newspaper enterprise or an editor. Besides, governments also provide other social amenities and services. This is the point made by a Ghanaian parliamentarian when he said:

The government has a duty to the people, and it is important that its views are clearly reported and not twisted in any way. For that reason, there is a case for the government having a press where its views will not be distorted. But that does not mean that the Government should monopolize the papers. All that I am saying is that it is important for the government to have an organ which will publish its views correctly (Ghana 1970).

But if, as has been demonstrated, the government has a right to own and operate newspapers, this right does not confer on it a right of monopoly in that domain. This is because the government-operated newspapers are likely to present only one viewpoint, and hence the need for other avenues for expressing other points of view. In the libertarian tradition, private ownership of newspapers is seen as an index of press freedom, but the question may be asked: Whose interests does the press represent and on what democratic basis can it claim to represent or reflect public opinion?

Since the general public has nothing to do with the appointment of editors and reporters who have the capacity to mould public opinion, it can be said that the private press represents class or sectional interests or the interests of its owners and nothing more. So what is the basis of their authority to inform and educate the public or to shape public opinion? In other words, what gives the press legitimacy as a kind of "Fourth estate?" This is the point made by the American politician, Rusk (1974: 18), when he said:

Let's get rid of this genial myth of the fourth estate. . . . That this should be so would seem to be elementary because the American people have nothing to say about who are to be publishers and editors and reporters and columnists. We cannot admit in our constitutional system room for something called a "fourth estate" which has no democratic base.

Echoing the same sentiment a few years later, but with specific reference to network television, former U.S. Vice-President Spiro Agnew asked:

What do Americans know of the men who wield power? Of the men who produce and direct network news — the nation knows practically nothing...is it not fair and relevant to question its concentration in the hands of a tiny and closed fraternity of privileged men, elected by no one and enjoying a monopoly sanctioned and licensed by government — it is time we questioned it in the hands of a small and unelected elite (Keogh 1972).

If such fundamental questions can be raised about the democratic base of the privately owned media in a country with a long tradition of almost exclusive private ownership of media organs and with well established constitutional guarantees, it should come as no surprise that in countries lacking the long tradition or experience of an independent, private press, the government's right to be involved in the operations of the media is almost taken for granted.

Whatever the system of media ownership in any given country, if the media are to serve both the ends of democracy and development, a certain amount of diversity is called for. Even though there is no consensus on the concept of a free press system, press freedom is measurable in terms of the extent to which different views, including the dissenting and unorthodox ones, are accommodated and given expression. It can, therefore, be asserted that a plurality of channels of information will serve the interests of democracy better than a monopolization of the channels by the government. An ideal mix would thus be a system in which government operated its channels while constitutionally guaranteeing the right of private entrepreneurs to set up their own newspapers to provide alternative sources of information.

It has been argued that providing alternative sources of information can pose problems for developing countries. The argument is that one major problem facing newly independent nations is that of achieving national integration and creating national consciousness that will distil local and tribal loyalties into national loyalty. To achieve this end, it is necessary to create national symbols through a controlled and centralized media system at the national level. It is further argued that given widespread illiteracy and inadequate political consciousness, a diversity of the sources of information or multiplicity of voices in the media can only create confusion in the minds of the people and thus render the task of nation building and development more difficult. It is this same type of argument that has been used to explain the creation of one-party political systems.

The argument sounds plausible but it is patronizing and assumes that either there is only one acceptable version of an issue or that the people are incapable of weighing alternatives and making reasonable choices. In order to educate people on their civil rights responsibilities and create in them the political consciousness that will enable them to participate meaningfully in the governmental process through periodic elections, they need access to information and they should have the right to all possible avenues for obtaining the necessary data to enable them to participate in public discussions and debates so as to influence decisions. Despite the need for the free flow of information in democratic societies, governments tend to consider that this entails risks because opposing views and dissent may be irresponsible and calculated to undermine stability. It is further felt that opposition elements may take advantage of the illiteracy of the masses and exploit their ignorance to advance their own cause. It is for this reason that governments see some political virtue in restricting the flow of information for the sake of national unity and stability.

In discussing the place and role of mass communication in the democratic process in Africa, it is important to recall the commonplace observation that the press system of any country is a reflection of social, political and economic environment in which it operates. Even though all countries lay claim to democratic ideals and practices, if we see democracy broadly as being characterized by participation and choice, it is obvious that very few countries on the African continent can be considered democratic: single-party regimes without any meaningful choice of leadership or policy alternatives, or military dictatorships dominate the political scene. In this kind of environment in which pluralism is not tolerated, it is unrealistic to expect a press system that constitutes a channel or vehicle for national debate. The fortunes of the mass media are closely linked with those of the political system of which it is part and, for that reason, it is pointless to isolate the issue of the mass media and discuss it outside the general political framework. Until the society becomes more open and tolerant, the media will continue to act as propaganda organs for the powerful elite rather than as public vehicles for rational discourse and debate on national issues.

It can be said that it is this absence of debate on national issues that has posed the greatest threat to political stability in Africa. In the absence of any recognized opposition that will propose alternative policies, the military has often seen itself as a legitimate opposition and has intervened in the political arena to redress what it perceives as the wrongs in the society. The intervention of the military hardly improves the situation: in many cases the regimes get more intolerant and the cycle of instability is resumed. This vicious circle can be broken only if there is genuine goodwill to establish open political systems in which free discussion will permit the weighing of alternatives and ensure that the government is continuously called to account for its performance, which brings the discussion to the watchdog function of the press.

Watchdog Role of the Press

One reason given by African political leaders for ensuring the control and subservience of the press as well as other media is that since the media are vital to the exercise of political power, their use should be closely controlled so that they are not used to propagate views and promote interests that are at variance with those defined by the national leadership. It is argued that since political institutions in developing countries are fragile and any criticism of the government may be interpreted as a challenge to the legitimacy of the government, the media should refrain from scrutinizing the affairs of the government too closely. The media should, therefore, confine themselves to serving as a one-way conveyer belt between the government and the people and helping to educate the people on the development goals defined by the government. Within this context, it is argued, the right to criticize as contained in the freedom of the press becomes a luxury which developing countries cannot afford. This means in effect that the media should have no watchdog role. How valid is this interpretation of the role of the press?

The concept of democracy is one which has a wide diversity of interpretations, but however it is defined, it implies accountability which is established through periodic elections even within one-party systems. But what happens between elections? How does government render account of its stewardship on a regular basis? If such accounting is necessary, what institutional framework can ensure this on a continuing basis? The obvious answer would seem to be the mass media which can exercise a regular scrutiny on the activities of the government to see how performance matches promises or how programmes are being implemented. Even in the African context where one is admittedly dealing with new and fragile political institutions, the press should act as watchdog of democracy. In a democratic society, actions of the government, which is only a trustee of the collective will power of the people, are expected to be regulated by the force of public opinion, and the press is the most appropriate medium for gauging and reflecting public opinion. In the absence of any such mechanism for regularly monitoring and evaluating the government's performance before the bar of public opinion, there is a great likelihood of the government falling into complacency, unresponsiveness and irresponsibility.

In this context, it is legitimate for the press to fulfil the role of an opposition in the sense of presenting another point of view where necessary; that is to say, criticizing government decisions which are not in the best interests of the people, denouncing abuses of power in society and defending human rights. A press or media system that decides to do less than this reneges on its responsibility and fails to contribute adequately to the democratic process or to national development based on democratic participation and decision-making. The press should not see its role simply as that of a deflecting mechanism, allowing people to let off steam as a kind of safety valve while the weakness criticized still goes on. It should be a substantial vehicle for reflecting the various shades of public opinion and articulating the people's feelings. The press in its watchdog role should, as a matter not only of right but also of duty, expose and criticize bureaucratic incompetence, corruption, abuse of power and the violation of human rights.

The African press played the role of a watchdog during the independence struggle and many of the nationalist leaders established newspapers for the purpose of organizing and mobilizing the people to fight against colonialism and injustice.[2] Some people who see this adversarial role of the colonial press to be justified now argue that with the attainment of independence, the press should play a supportive rather than a critical role. This is because the governments need the support and sympathy of all to tackle the gigantic problems of development. Those who argue along these lines fail to ask a fundamental question: What were the evils that the press in colonial Africa fought against? The press fought against political oppression, economic exploitation, social injustice and the abuse of human rights. A further question to ask is: If the evils denounced by the press still persist, and the press proved to be an effective weapon in fighting against those evils, why should it not continue to be used in fighting a fresh manifestation of those evils? Should the press abdicate its responsibility of fighting and denouncing certain evils simply because the perpetrators and victims happen to be the same colour?

Press freedom is not absolute in any political system, but this fact has often constituted a pretext for denying it altogether in many developing countries. While it is reasonable to argue that a certain amount of restraint and moderation is needed in developing countries to protect and help stabilize the new fragile political institutions, the press will be abdicating its important responsibility of watchdog if it fails to constantly scrutinize the government's activities.

Summary and Conclusion

The media in Africa can serve the ends of both democracy and development. Even though there is no tradition of the Western press in pre-colonial African society to draw upon, there are sufficient elements in the African conception of human rights to provide a solid base for a press system that tends towards, or is inspired by, liberalism rather than authoritarianism. If there appears to be disregard for human rights in Africa, the reasons may be sought

elsewhere other than in African traditional, political and cultural practices.
Contrary to the assumption that, at this stage of socio-political evolution in Africa, development and democracy are mutually exclusive, the two concepts are reconcilable and can be pursued concurrently. In fact, it can be argued that the concept of development as redefined by Rogers presupposes the existence of certain basic democratic principles. The notion of participation which is an essential ingredient of the concept also presupposes access to information by the citizens to enable them to participate meaningfully in public discussion and decision-making. It is an outmoded, mechanistic concept of development that sees the freedom of expression as a hindrance to its achievement.

It is generally agreed that in addition to the traditional functions of the press, namely, informing, educating and entertaining, the media of developing countries have the additional responsibility of promoting development. The problem arises when discussing how this developmental function may be carried out. While some argue that the press can play this developmental role only when it is subjected to authoritarian controls and direction, authoritarian controls tend to inhibit public discussion and restrict the free flow of information, thus eventually stifling and negating efforts at genuine human development.

Since the press or media system is part of the political structure of a society and operates within certain ideological parameters, any meaningful discussion of the role of the press must take into consideration the political philosophy of that society. An objective look at the African political scene confronts one with an environment that is not conducive to the pursuit of democratic ideals. It follows from this that any meaningful search for the African media in the democratic process should be broadened to include a search for those political institutions that will enable the media to be both a tool for social transformation and a watchdog over the people's rights; in other words, to serve the interests of both development and democracy concurrently.

NOTES

1. This is an edited version of a paper of the same title presented at the 6th Biennial Conference of the African Council on Communication Education held 24–29 October 1988 at Jos, Plateau State, Nigeria.

2. One can cite as examples such papers as Kwame Nkrumah's *Evening News*, Jomo Kenyatta's *Muigwithania*, Nnamdi Azikiwe's *West African Pilot*, Herbert Macaulay's Lagos *Daily News*, among others.

Chapter 3

A TAXONOMY OF TRADITIONAL MEDIA IN AFRICA

Des Wilson

Introduction

This chapter draws largely from field data gathered from the south eastern Nigerian states of Akwa Ibom, and Cross River. However, there are similarities which cut across all of Nigeria, Cameroon and the west coast of Africa with some slight variations in east and south Africa. The prevalent mode in a particular region is largely dependent on the fauna and flora of the area as well as mineral and other resources available. These are essential to the manufacture and construction of communication devices as are the relevant skills needed for their production.

The various modes of communication are broadly divided into six classes. The classification may be contested by other scholars. However, one will contend that they at least provide approximations which can prove useful in the quest for an understanding of indigenous communication systems in Africa. The six classes are: instrumental media, demonstrative media, iconographic media, visual media, extra-mundane media and institutional media. These are in turn subdivided into sub-classes of media instruments which are as diverse as there are villages and clans in Africa. These classes and sub-classes are discussed fully in the following pages.

Instrumental Media of Communication

Instrumental modes of communication include idiophones, aerophones and membranophones. These are instruments or media which when beaten, blown or scratched produce diverse sounds and messages based on the expertise of the traditional newsperson and on the nature of his or her message.

Idiophonic Communication Instruments

Idiophonic communication involves the use of instruments which are self-sounding. Such instruments include the metal gong, woodblock, wooden drum, bell and rattle. These instruments are capable of producing their own messages as well as producing signals which serve as attention-directing devices prior to the delivery of the actual communication message.

Idiophonic communication devices that were identified during this study included the wooden drum, the metal gong, and the bell.

The Wooden Drum

Among the most important of the idiophones is the wooden drum, which is known in south eastern Nigeria as *obodom* (wooden drum). The people have a saying that the *obodom* is the message-bearer of the ethnic group. It is a very common instrument in most of the southern states of Nigeria especially in the states across the River Niger to the east. It speaks the drummed language. Most African people use it both as a musical instrument and as a medium of communication. Akpabot (1975) describes the wooden drum as a

> drum, a hollowed out tree trunk made to produce two tones. It comes in different sizes and it is played with two beaters made out of bamboo. . . . It can be played as a solo instrument by a specialist musician to transmit messages from the chief of the village, in groups of two or three in an orchestra.

Akpabot (1975) adds that it comes in various sizes and performs different functions. Among the Ibibio of south eastern Nigeria three major kinds are identifiable namely, *obodom ubong* (royal drum), *obodom mbre* (common drum used by masquerade groups), and *obodom usuan etop* or *obodom ikot* (drum for message dissemination). *Obodom ubong* is a two-piece medium for disseminating information that has a direct link with royalty. One of these drums is smaller than the other. It is known to be used on three specific occasions, namely, at the installation of kings, a royal celebration and at the death of kings. The language of the *obodom ubong* which is based on the tonal patterns of the local language is understood by those who have been brought up under the traditional system. As it is usually played by specialist drummers, its language is not readily accessible to most members of the younger generation. To make this possible, the present generation has to be taught its system just as they learn the letters of the alphabet.

The *obodom mbre* consists of *obodom idion* (*idion* cult drum), *obodom ekpo nyoho* (*ekpe* masquerade drum), *obodom ekong* (*ekong* society drum) and others named after the cultural groups which make use of them. These drums are also carved out of logs of wood. Omibiyi (1977: 25) says this of them:

> These are hollowed out and slit open at the top to create a pair of lips which are struck with beaters. The two lips give contrasting tones. . . . They are used to accompany various music and for verbal communication.

Obodom drum messages are addressed to specific individuals by calling their names and summoning them to the chief's home, or the whole community may be so informed through this medium if the message is meant for public consumption. This very vital function is performed by the *obodom usuan etop*. By its very name the *obodom usuan etop* was meant for sending messages. It is also the *obodom ikot* which was mentioned earlier. Some informants reported that because this type of *obodom* is very large its message could cover distances of over twenty kilometres. Chief William Ufot of Ete in Ikot Abasi (Nigeria) explained that when played at night its message could be received in very distant places. In some cases, he claimed, the message can only be understood by the elders and chiefs for whom the message is usually meant; and that in the past the *obodom* could be used to invite everyone to the village square within a very short spell of time. Then at the village square the message was delivered to everyone in a manner similar to what obtains at today's mass rallies.

The *obodom* used in the different masquerade groups produce esoteric messages meant for the ears of members. For example, only members of the Ekpo Nyoho society understand the speech formula used to deliver special messages to its members. The playing of the *obodom* to produce messages is an intricate affair which demands a high level of professionalism and expertise and such is found among those who have been trained to play it. No outsider nor one without prior training is allowed to do so.

In Mbarakom, Akamkpa Local Government Area, the *Obodom* plays out a call signal to farmers in the bush and this hurries them home to hear the message behind the urgent call of the village elders and chiefs. The first port of call of the home-coming farmers is usually the home of the village head or other chiefs. An informant reported an experience in which villagers ran into the forest when Federal troops entered the town during the Nigerian Civil War, and all appeals to them to come out and return to their homes failed. It was not until the village head ordered the playing of the *obodom* that the citizens mustered up courage to come out, thus giving credence to the claim that traditional media of communication have credibility and reliability.

In Ikot Obong village in Ikot Abasi Local Government Area, (Nigeria), Chief Ben Ikwot also spoke of the significance of the *obodom ubong* of which he said, when the citizens heard its sound they would exclaim: *Iya nkpo atibe!* (Ah! something has happened) and would then hurry to the village head's residence or to the *efe* (traditional shrine or traditional meeting hall) where they were told of the events that led to the call.

Thus, apart from being used as musical instruments, the *obodom* is also used in rural areas for the dissemination of information. Communication through the *obodom* is two-way. The audience responds to its messages. The response, which may come from an individual or a group in the community is always simultaneous and spontaneous.

From the evidence adduced from the field interviews, the following were observed as the functions of the obodom:

i.   It is used at the installation of kings.
ii. It is used at royal celebrations.
iii. It announces the passing away of kings.
iv. It informs the citizens of grave danger.
v.   It is used in various masquerade groups, namely; *ukwa, ekombi, ekong, ekpo nyoho, ekpe* and many other musical entertainment activities.

Metal Gong (Nkwong)

Akpabot (1975:15) described the character and functions of the gong (*nkwong*), which shares a lot of characteristics with the Akan *dawuro* described by Ansu-Kyeremeh, as such:

> Next to the drum the gong is the most frequently used in Ibibio music. It is very prominent in the music of secret societies, where it can be used as a solo instrument to announce the impending appearance of a masquerader . . .

According to Akpabot (1975), there are three types of instruments by the generic name, gong. Even though they are all closely related, the *nkwong* (metal gong) which is the largest of them all, is also the most frequently used. It is used in most female societies both as a musical instrument and a medium cf communication especially in announcing dates and times of meetings and festivals.

The *nkwong* is a large conical metal gong which is normally beaten with special sticks or carved bamboo rods. It belongs principally to the Ebre Society, which shares the characteristics of the precursor to the Mityana women's clubs discussed by Nabasuta Mugambi. The *nkwong* is meant to be played on very important festive occasions and during the death of one of the female members of the society or an old, dignified, prominent and reputable woman who on account of old age no longer was able to participate actively in the activities of the society. In the earlier days whenever it was played it was an indication that something of importance had happened to a member of the Ebre society. Its sound served as a rallying point for all other women in the village.

In other circumstances, the smaller gong, the *ekere* (a ritual gong), is used to announce the death of an old reputable woman. The Ebre Society (a cultural association of virtuous women) also has an appointed messenger whose duty it is to go round the streets and paths of the village mainly in the evenings and in the early hours of the morning to announce the news of an event that had either taken place or was about to take place. It is important to note that whatever announcement or information the women have to make or share does not affect the men or women who do not belong to the society. It is easy for rural dwellers to determine from the medium used and the voice of the "broadcaster" whether the message is meant for all the citizens or not.

The announcer, who must be a female, beats the *nkwong* as she goes round the village delivering her message. She does this with a stick similar to the one used along with the *obodom*. The sound of the instrument attracts attention to her person. She then announces the message by reciting whatever information, news or instructions to the womenfolk who would constitute the target audience.

Esen (1982: 127) writing of the royal significance of the *ekere* and its use as a musical instrument says it "is used mainly by the Ibibio chiefs on very special festival or ceremonial occasions." It is important to note that some of the instruments which have so far been discussed had varied uses to which they were put, each locality determining the expedient function of each instrument. Thus, when Esen (1982) says that the *ekere* is used principally by Ibibio chiefs, such a generalization should be seen within a specific social context about which he wrote. Esen (1982: 127) also says about the *ekere*:

> Its sound, smooth and mellifluous, denotes joy, victory and the gladness of success. When a chief is installed he sounds the ekere as a symbol of the joy of achievement.

Other uses to which it is put are listed by Esen as including the celebration of war victory by war chiefs, and eulogy at a public oration on the passing away of a hero. And he sums up his views thus:

The *ekere* is therefore in a very real sense the musical instrument of Ibibio royalty and priesthood.

The *nkwong* is also known to be used to spread news that is of minor importance especially that which involves social clubs. The double-bell shape of the gong variety used for this purpose makes it possible for it to produce two tones thus making it an appropriate musical instrument.

Traditional rulers in the area have identified the following as functions of the two main varieties of the gong used in the area:

i.   It is used to speak to ancestor-spirits during *idiong¹* perfor-mances.

ii. Female societies, especially in Ebre society, use it for dissemi-nating information, news, and messages related to other social activities.

iii. The *ekere* is used by men in Ekpo society.

iv. It is used to call the dead during one's funeral.

v.   The *akpan ikpo* (the successor to the deceased king) holds it at his coronation.

vi. It is used by chiefs at their installations as a symbol of the joy of achievement.

vii. It is the musical instrument of Ibibio royalty and priesthood.

Woodblock (Ntakrok)

The *ntakrok* (woodblock) was primarily used as a musical instrument but it is also used for minor information dissemination duties. It is made from wood and is hollowed inside but flat on its sides and it is played with a stick. It is used in much the same way as a metal gong. It is used to announce community projects, to place injunctions against the harvest of economic trees in the village, and also used by family groups to announce their programmes.

Bell (Nkanika)

The bell has a way of forcing its presence on the audience. It cannot be said to be typical of any particular society since it is present in almost any known part of the world in its various shapes and sizes and also performing closely related functions. In Old Calabar traditional society, the bell at tennis strongly associated with secret societies like *Ekpe* (a male-dominated cultural society which also performs judicial functions of the type performed by judges and magistrates), *Ekpo Nyoho* and *Ekong* societies. The bell today transcends this function of forcing itself on the people and reminding them of the presence of any of the traditional masquerades mentioned above through tintinnabulation. Today it is generally used at social gatherings to draw the attention of the participants to an on-going activity. It is also used by itinerant advertisers in rural and urban areas to attract the attention of the passers-by to the wares that are being sold. It is frequently used at bazaars in conducting sales.

With regard to its effectiveness as a medium of communication it has a great potential which has perhaps not been fully tapped especially when the communicator rings it round the village on a bicycle or motorcycle or even on foot. Just as the other media already mentioned are used in communication as attention-directing devices, the bell also serves this purpose.

Aerophonic Communication Instruments

Aerophonic communication involves the use of instruments which produce sound as a result of the vibration of a column of air. The sounds may be messages on their own or they may serve as signals. The instruments used for this purpose include flutes and horns from plants and animals, and whistles. Some of the most popular ones are cow, ivory and deer horns.

Ivory Horn (Nnuk Enin)

This medium is regarded in most parts of Ibibio land as the most dignified and royal medium of communication. It has a sacred, definitive function in traditional communication. Its importance also lies in the fact that it is used in rare circumstances. Unlike those media which speak in coded "languages" its messages are simple enough to be easily understood and related to events that are usually associated with their use. Again, this can be likened to the Akan *abentia* (short horn) which Ansu-Kyeremeh describes later. They are both made from elephant tusk and are elitist, exclusively owned by chiefs.

This medium is made out of the tusk of an elephant and it has a hole made at its tail-end through which the message is blown. This is one medium that is not handled by the ordinary messenger of society unless one's role as communicator has been enhanced by that individual's personal status in the society. Its message could be cryptic and also meant for the masses of the people. It plays a very vital role in rural society. It is used as an instrument of peace and also used for information dissemination purposes. Under these broad roles the following specific functions have been identified:

    i. It is blown to settle quarrels.
    ii. It is used on very important occasions to inform citizens of the death of kings, serious calamities and other grave occurrences.
    iii. It is used for placing injunctions over disputed land or property.
    iv. It offers the final word or judgement on issues.

v.   It is used by secret societies to inform members about important festivities like offerings to the deity.

In times of war it is blown to alert the people and also to mobilize the young men and women in society. On such an occasion the blower climbs to the top of a tree or house to make its sound get to very long distances.

In cases involving communal or ritual killings as was the case with the dreaded *Ekpe Ikpaukot*, the chiefs and elders of the society took the elephant tusk and stood at the village centre of the erring group and blew a warning message against a repeat of such murders which occasioned the activities of the *Ekpe Ikpaukot*, otherwise known as "Man Leopard." This action of the elders meant that an injunction had been placed over the community. When they returned to their homesteads they were somewhat sure that those rituals would not be repeated. But in cases where the breach of the injunction occurred, Chief William Ufot of Ete in Ikot Abasi asserted, a second visit to the erring community led to the land being turned to scorched earth. Wives and children of the recalcitrant community were also taken away and shared. All these gave the tusk the image of a final arbiter in all matters of state. In a sense, it is an example of a McLuhanite medium-derived message. Today, the tusk still retains some of its aura of importance but the convention and circumstance that make it the final arbiter are no longer in existence, yet it is still regarded as a powerful instrument or medium of communication especially for its rarity.

Perhaps the importance of the tusk in the past and today is borne out of the fact that it is a highly-prized and cherished economic product and hence it is only found in the homes of kings, chiefs and the rich. Talbot (1923) says Afaha Eket Chiefs in Eket exacted it as tribute and royalty from the early Ibeno settlers until the practice was abolished in the late nineteenth century by the early British Missionaries.

Today the tusk no longer commands the same royal dignity that was accorded it in the past but it still retains its royal significance on the death of kings. When it is blown, the citizens know that something serious is amiss. This symbolism is not totally lost on the people.

Wooden Flute (Uta)

The *uta* is a conical instrument made from a special kind of gourd whose natural shape makes it possible for the instrument to be made and used for the purpose for which it is now famous. The *uta* instruments are of different sizes. Akpabot (1975) distinguishes four different sizes but this excludes the very small *uta* which is of the size of the mouth organ. *Uta* is also the name given to a dance troupe which uses this instrument as the predominant musical instrument. It does not seem to have been originally fashioned out as a communication instrument and was only discovered for this purpose during its "talking" sessions in music. Its use was not visible during the time of the investigations in the villages except during musical entertainment. However, oral evidence showed that wherever the *uta* was blown in the village, citizens trooped out to listen to the message which was intended for them. Chief Japhet A. Udoh of Ikot Ekpaw, the paramount ruler of Ikot Abasi, noted that the *uta* was never played except at very important events like the death of a "good" old woman. He also asserted that such a woman must have been a member of the Ebre society and that the Efik-Ibibio expression: *Men uta for kefit ko* means "go and blow your news elsewhere." This interpretation if valid gives an indication as to the communicative use of the *uta*.

Sam Akpabot (1975) adds however that an identical expression which he recorded during his field studies: *"Men uta for kefit ke Annang"* (Go and blow your *uta* in the Annang country) refers to the cultural origin of the *uta* itself as an important medium either for music or for communication. But while Obong Japhet Udoh saw the *uta* as a symbolic referent or synonym for news, Akpabot (1975) saw it as a representation of the medium itself. If the Annang origin is true, certainly the expression does not in any way make this clear as there is a strong suspicion that this could have been a derisive parlance at a particular time as is common in the language with all such expressions beginning with *"Men . . ."*

Nevertheless, the *uta* is a symbolic medium of communication in the sense that it conveys only one meaning, and that is, it calls the attention of the people to an on-going event or that which has occurred. Such an event could be the death of an important lady of the community or in some instances it could be an indication of some revelry going on in the house of an important citizen. In these circumstances the blower first sounds the *uta* and then announces the news.

In spite of its communicative capabilities the *uta* was not generally accepted as an important medium of communication. It was clear from the evidence that it was not generally acknowledged for this purpose. It was actually found to be of limited use to some communities. Its unpopularity as a communication medium is perhaps embedded in the apparent irritation in the remark *"Men uta of kefit ko"* (or *". . . ke Annang"*) which is also an indication that it may have been employed by some social clubs to carry out their communication activities in the past and through overuse it acquired its seeming notoriety and irritability.

Finally, the following can be said of the *uta*:

i.   It is a cone-shaped medium of communication made from the gourd, slit to enable the blower "play" whatever message is intended.
ii. It is now usually used at public festivals and rallies to sing praises about the leader or the event being celebrated.
iii. It is played for deceased old ladies usually members of the Ebre Society. In this respect it is never played for deceased young ladies.
iv. Social clubs use it to inform their members of activities.

Thus, the *uta* is significant in many respects as an instrument of communication even though it has not attained the same level of popularity as the other more commonly used ones.

The Obukpon
The *Obukpon* is made from the deer horn and it is used practically in the same manner as the elephant tusk and *uta*. Significantly, one of the early newspapers published in the Old Calabar Province took its name from this medium. It was called *Obukpon Efik* and belonged to the Presbyterian Mission. Its use could be equated more with that of the *uta* rather than with the elephant tusk, although Edidem Atakpa (1981)[2] claimed that *obukpon* is sometimes used as the symbol of the Supreme Being and that when blown the rainbow appears; and this, he said, is done twice in a year. *Obukpon* as an instrument of communication is mainly used as an attention-directing device and its prominence as enunciated in the Itu area was not supported by other areas. It became clear that its importance and function have been eroded through the use of other local instruments.

Membranophonic Communication Instruments

The membranophonic communication instrument is that which uses skin or leather drums to produce signals and messages through the vibration of the membrane which is beaten by hand or struck with a stick. They serve as signal source and produce messages of their own. The skin drums are found in all Nigerian, and indeed African, cultures and they come in different shapes and sizes.

In the location of this study they were known by various descriptive labels which tended to define the functions they performed for the various groups. They were also seen as musical instruments. Their descriptive labels actually derived from their musical function. Portability was important if they were to perform communication functions. The ones used for the purpose of disseminating information were usually small in size to enable the newspeople to carry them about. They are generally known by the name *Ibit*. Some call them *ekomo*. In the Ikot Abasi area the ones used in information dissemination matters are called *nkom* while a similar drum in the Onna area of Eket is called *ikpeti*.

A specific drum is beaten by the traditional newsperson as s/he goes round the community delivering whatever message s/he has. As information on the second question schedule showed, this drum is the most used of the various media of communication in rural communities.

Many respondents (65 percent) identified the skin drum as the most frequently used instrument of communication in rural areas. The traditional newsperson who is always known in the community plays it as s/he delivers or broadcasts his/her message. S/he has always been a kin relation of the community whom the citizens or receivers of his/her message can relate to and do relate to. His/her credibility is instantaneously verifiable. The audience can easily confirm or dispute (if need be) the information. This factor is probably what Schramm (1963) alludes to when he wrote that:

> At present among many transitional people there is still a strong tendency to appraise the reliability of various media mainly on the basis of the strength of their personal relationship with the source of information.

The skin drum is thus a very important instrument of communication in rural societies.

Symbolographic Communication Instruments
Symbolographic communication instruments entail the use of cryptic representation in the form of writing (pseudo-writing) made on surfaces — hard or soft — like the rind of the bamboo, walls, cloth or the ground and sometimes these may be in the form of signs. This is symbolic writing or representation which may be employed in communication among members of an exclusive club.

Nsibidi Writing
The best known of the symbolographic form of communication is the *nsibidi* writing of the Efik-Ibibio, Ekoi, and Ejagham people of the former Cross River State. This form of writing is also found among the Mom of the Cameroon, the Vai, Basa, Mende and Kpele in Sierra Leone and Liberia;

and also among the Igbo of Eastern Nigeria who are the neighbours of the Old Calabar people. As Alexandre (1972: 110) pointed out,

> The Nsibidi writing . . . seems to be in a transition phase, its character sometimes representing a complex notion and sometimes a sound.

As a writing form, it is not well-known on account of the restrictive usages among members of a secret cult which "uses it both to enable its members to communicate among themselves and to fashion amulets and magic charms." But among the Igbo it has been used to express more modernistic ideas by avant-garde painters and artists. *Nsibidi* writing is closely associated with Ekpe cult and represents perhaps what would have been one of the earliest forms of writing in these parts of Africa. But because of the secrecy which surrounds most of its activities Ekpe society members left this potential writing form undeveloped outside the confines of secret society esotericism.

Bamboo Writing (Nsadang)

This is a small, decorated stick made from the dry branch of the raffia palm tree. It is often marked to represent different codes. It has a limited usage although its specific use in times of disputes could spark off fratricidal conflicts, and intercommunal or interpersonal violence. Aspects of its secret codes are lost and perhaps this accounts for its indiscriminate use by individuals in certain rural societies today. Chief Anangabo Effiong of Mbarakom Clan in Akamkpa Local Government Area was of the view that it used to be a powerful instrument of communication in his area in the past but that it has been replaced today by other instruments. Slave messengers, claimed the Chief, conveyed the *nsadang* from one chief to another when the messages were destined for distant places. Although other interviewees disputed the use of slave messengers over long distances because of the possibility of their escape, others insisted that such slave messengers would have lived long enough in the host areas to have planted roots through marriage and the bearing of offspring and could thus be trusted not to escape. This, in fact, is the view of Noah (1980: 17).

Whatever was the case, slave messengers were a rarity when the message involved a long distance. Nevertheless, the notion of slave messengers still suggests humble beginnings for an information dissemination vocation and not necessarily a lowly profession. Its beginnings were simply humble. On the whole, it was clear that only trusted members of the society were made message bearers in traditional societies, and this pattern exists till today even though traditional practitioners may be fewer.

In a different sense, when the cryptographic marks on the *nsadang* are considered, it does not seem as if it commanded the same mass function as other media instruments which have been discussed so far. Perhaps it could be said that the information received through the *nsadang* prompted the use of another medium to amplify it. Besides, the *nsadang* does not have the information capacity of saying much beyond the cryptic marks made on it and this language is clearly understood by both the sender and the receiver. In spite of this, the bearer sometimes had to say a few words in explaining or introducing the mission. Whatever its merits then, *nsadang* combined the written and spoken word; it was a form of writing that was supplemented with verbal communication in its multimedia format.

Today *nsadang* can perform the function of private communication between leaders as is often found in diplomatic circles. It is a powerful medium for delivering messages from one leader in the village to another, or from one family to another. Even though it has been trivialized by the frequent recourse to its use by people in conflict, it still maintains its potency. Thus today, the following functions are still discernible:

i.   declaring disputes between two villages, clans or families.
ii. summoning parties in disputes just as *eyei*, another potent medium of communication which belongs to another category of media.
iii. indicating a warning by those who feel offended by others.

Its sharp end is said to be indicative of the acuteness of the problem in relation to which it is used. When both ends are sharpened it is said to represent a very serious quarrel which may be difficult to resolve and when one end is sharpened it is said to indicate the possibility that though the quarrel may be serious there is still room for an amicable settlement.

Demonstrative Communication

This form of indigenous African communication is more of an aural communication. It uses music and signal as modes of communication. Many of the media already discussed in this chapter are used as musical instruments, and music in itself is a mode of group and mass communication in all societies. Music, as Jacobson (1969: 334) pointed out, is "an unconsummated symbol which evokes connotation and various articulation, yet is not really defined." From the inaudible music of the spheres to today's high fidelity stereophonic systems, music has played a significant role in ordering, re-ordering and generally shaping human society. Dietz and Olatunji (1965: 1) noted that

> music is part of everyday work, religion, and ceremonies of all sorts. It is . . . used for communication. Many tribes have no written language, so they send messages by word of mouth through singing, blowing signal whistles, or by talking drums which imitate the pitch of the human voice.

In a similar tone, Ekwueme (1983: 4) stated that "Functionality is a known feature of music and the arts in Africa" and that,

> In this functionality, communication becomes a primary or at least secondary objective.

African music, in the Hornbostelian sense is the 'life of a living spirit working within those who dance and sing" and this view readily applied to its function as a means of communication. Besides, Akpabot (1981) sees such music as a vehicle of social change capable of being used for religious change, social order change and educational expansion. Euba (1982) also agrees that music can be used as an instrument of social change.

In some traditional societies, grapevine stories are presented in songs by cultural groups and other social groups. Among the people of the Old Calabar, for example, itinerant music entertainment groups use satire, criticism, moralization, praise, symbolism, didactism, suggestion, labelling and name-calling to communicate with individuals, groups and the society at large. Such groups may include the *akata, ekpo, itembe* and age grades. Current events (gossips) are presented in their lurid details especially when they are about the rich and proud. Today, the phonograph record and cassette tapes, mostly through radio, television and video have contributed to make music a mass medium of communication, whether in its recorded form or in a live performance. And music has become a powerful instrument for interpersonal, extramundane, intra-personal and mass communication.

Instrumental modes of communication thus produce sounds that signify or symbolise a communication event within the context of a specific setting. The sounds of the different drums, flutes, horns, bells and gongs serve as signals of communication. Signals are often accompanied by oral or visual messages depending on the communication context.

Iconographic Communication

Iconographic communication involves the communication of ideas or information through the use of objective or concrete reality in inanimate or animate form. The two main modes employed in this category are objectified and floral communication.

Objectified Communication

Objectified communication (or objectics) is a signification in which the object refers to a thing, event or concept. This concrete representation may have a limited meaning or may have a universal application or significance. The presentation of a bowl or saucer of kolanuts to a visitor has significance within the context of the presentation and also has symbolic meanings. Such meanings are more powerful than what words could ever convey. The same goes for the presentation or exchange of charcoal, white pigeon or fowl, white egg, feather, cowries, mimosa, flowers, sculptures, pictures, or even flags in modern ceremonies.

Floral Communication Instruments

Floral communication involves the use of selected flora of the local vegetation for the purpose of communicating specific meanings or ideas to members resident in or who may pass by the particular place where these media are used. They often act as traditional billboards. The most common ones used among the Old Calabar people include the boundary tree, a species of elephant grass, mimosa, *eyei*, *nyama*, *isara* and flowers. For example, the plant usually used to demarcate boundaries between plots conveys strong cultural meanings in relation to land matters. In addition, the young unopened frond of the palm tree (*eyei*), referred to by the Ibibio as *nwed Ikpaisong* is used in many ways to convey various meanings and it also performs various functions.

Young Palm Frond (Eyei)

The *eyei* is one of the most important media of communication of the past which still retains its potency and effectiveness. Johnson (1932) as district officer in Abak wrote about clan meetings which were always convened by the Clan Ayaraufot, who sent round *eyei*, a fringe of palm leaf, to every village as a sign of authority for the messengers who delivered notice of the meetings.

Two things come out clearly here; the first being that the *eyei* is a symbol of authority; secondly, it could not be issued by anyone else except the village or Clan head. It is still considered a symbol of great importance and urgency whenever it is used to communicate a message publicly or privately. Akpabot (1975: 22) offers some explanation in connection with its use in Ekpo Society (a predominantly male cultural group which also performs some judicial functions), which in the past and to a lesser degree today, was considered part of the traditional judicial system, adding that whenever,

> the chief of a village did not want a piece of land cultivated for some time, he would send out a musician of the *Ekpo* society who would go round the village playing an *Ekpo* drum (*ibid ekpo*) to announce the chief's decision.

This action was then followed by another symbolic announcement which in this case was in the form of "a branch from a palm tree (*eyei* . . .) no one dared cross that mark." And if anyone defied this mark or symbol of traditional authority, Akpabot (1975) says "he would be found dead under mysterious circumstances, a victim of the ancestral spirits." Whether in fact this happened or is capable of happening is one of the assumptions that are couched in traditional myths and beliefs and may be extremely difficult to prove. But one thing is clear and that is the authority of the *eyei* which comes from the village head has some divine backing and if the chief in his human manifestation cannot punish the offender against tradition, tradition alleges that the supernatural forces who are in communion with humans punish such a person in whatever way they deem fit. This notion of divine intervention in human affairs and communication between humans and deities is a popular view in sub-Saharan African cosmology. For, as Wilson (1975: 6) points out in another context,

> There is territorial contiguity . . . one world merges into the other like the upper and middle streams of a river.

This world view is still very strong and is responsible for the level of social stability in the society, although some have mistaken this fear of powerful and unknown spirits for illiteracy, ignorance and superstition. Today, it is not uncommon for the poor illiterate supplicant and the western-educated Christian priest to meet at the traditional soothsayer's shrine seeking protection from the powerful spirits which they fear their enemies are about to unleash on them.

Some of the interviewees in Eket and Ikot Abasi spoke of how federal legislators and other important sons were invited to crucial meetings by enclosing *eyei* in their letters of invitation. Some considered the enclosure of *eyei* as the strongest factor in getting people to respond to such invitations since previous open invitations were ignored. Perhaps it can be surmised here that a fear not unrelated to that of the supernatural forces in the symbolic presentation of *eyei* may have propelled the invitees to attend those meetings.

Yet *eyei* on its own does not convey any specific meaning but acquires meaning within the context it is used. For an example, an *eyei* presented to a suspected witch or wizard is a sign of notice of ostracism. An *eyei* hung around a disputed land is an indication that an injunction has been placed on the use of the land until the dispute is settled. *Eyei* has religious as well as judicial functions. In its communicative function it is a sign of peace.

In rural areas where farmlands still constitute potential trouble spots, *eyei* is strung around disputed farmlands. Passers-by understand the message whenever they see it. Although the message is limited to only those who pass by this does not in any way reduce or diminish its effectiveness, because it is meant to speak to those who pass through the area. As many as pass by receive the silent message loud and clear. *Eyei* is never issued by just anybody or without the authority of the head of the village or community. The head of the family may also, in some special cases, use the *eyei* for special effects in matters involving two or more families.

One important advantage this form of communication has is that, like Western modes of communication, its message can be amplified through other media or channels of communication for a wider coverage.

In sum, the following general functions of the *eyei* can be identified:

i. It can be used to ban offenders from further participation in community affairs, or in preventing them from going out until they are found guilty or guiltless. The traditional judicial system is such that the accused is adjudged guilty until he proves himself innocent.

ii. It notifies people of the presence of a shrine in a particular area and non-indigens and non-initiates are usually expected to keep off such shrines.

iii. It also notifies the general public about certain routes they are expected to keep off if members of a secret cult were going to use them for a specific purpose and within a specified period. This kind of restriction is partly responsible for some of the clashes non-indigens and iconoclasts often have with traditional society.

iv. When displayed around a piece of land, it serves as a warning over the use of such a land whether in dispute or not.

v. It is used for arbitration in times of conflict and is thus used to restrain factions from continuing in their feuds.

vi. It is used to ostracise witches and wizards and other persons considered to be engaged in acts detrimental to the progress of society.

vii. It is used to indicate the importance of the event the receiver is being called upon to attend.

viii. It is used when one person is seeking assistance from another in a private matter.

ix. It is sent by kings to fellow kings or to vassals to point to the authority behind the message and the urgency with which it should be treated.

x. When issued by the crown it is delivered through a messenger, and the *eyei* is tied to indicate the message. In this form it resembles and imitates writing and it is thus regarded by traditional rulers and elders as the universal newspaper or literally "the book of the land" or *nwed ikpaisong*.

xi. During celebrations they are used as decor. Their mere presence conjures an atmosphere of solemnity or celebration depending on the time of year or occasion for which the palm fronds, including the more mature ones, are displayed.

xii. They serve as notices that a corpse is being conveyed in a vehicular procession. They are usually tied to such vehicles. A variety of other leaves are used by other societies for this same purpose.

xiii. They are used to warn against certain anti-social activities of individuals. In such instances they are presented to those who may be involved in such activities and they are either summoned or verbal warnings accompany their presentation.

Visual Communication

Visual communication is what happens to the receiver when a certain effect is produced by light on the eye and this brings about his realization of the different shades of colours as may be visible through a colour spectrum. This form of communication also involves the changes which the differences in the form of clothing, appearance and general comportment on the part of the communicator produce on the receiver. This sensation caused by light makes it possible for a person to carry out a differentiation of colours and attach his meaning and symbolism to them in his cultural context. The individual receiver of visual communication has the advantage of utilizing the two major advantages of pictorial communication. These advantages lie in the speed of the impact of the message and the freedom of the visual information from linguistic barriers associated often with written and spoken language.

The use of colours for the purpose of communication has remained for a long time a very important mode of human communication. All societies have had a relatively common symbolism attached to the different colours produced by human beings or present in nature. But while others have retained theirs some have, through cultural imperialism, been deprived of their original notions or symbolisms. Racism and ethnocentrism have tended to obliterate the culture of many technologically-deprived nations.

In Western societies for example, light colours are associated with pleasure, and dark colours with sadness. Some of the racially biased anthropological reports like that of Turner (1969: 59–62) speak of "white symbolising milk, purity, health, [and] good luck, while black represents faeces and other grim things." This archetypal racist notion is derived from the Middle Ages where black was suggestive of material darkness and was symbolical of the spiritual darkness of the soul. Later, it was, as Hulme (1899: 28–29) points out, incorporated into ideas of the Devil — the Prince of Darkness, and witchcraft — the black art.

Yet in spite of this corruption of values and colour systems of some nations, the seemingly similar symbolic colour systems cannot be adduced simply on the basis of the existence of colour universals. Among the people of Old Calabar Province in general and the Ibibio in particular, there is some general display of agreement in the use of the following colours individually or in combination to depict certain cultural events, taboos or feelings.

Red expresses danger and spirituality. It is a favourite colour of the secret societies, namely, *idiong* and *ekpe*. It symbolises fire and blood. White also expresses spirituality and purity. At times it is used as a symbol of mourning in memorial rites pertaining to dead elderly persons or in similar rites within christian religious contexts. It is a colour also used in shrines and other religious institutions. Black is a favourite colour of the Ekpo society as well as Ekong and other masquerade groups used ostensibly to depict the fact that these are spirits or ghosts. It is also used to represent the dead as well as mourning. It is perhaps the most abused colour in racial disputes. Brown is the colour of the earth and it is used in "repainting" of homes during major festivals or periods of festivals. In rural areas most houses wear the brown colour because it is the colour of the soil that is most used in so-called mud houses. Green is also the colour of Nature. The verdant leaves of plants are used in masquerade groups and for decoration during festivals and ceremonies. It is a colour representing our umbilical link with other things of nature (plants) as well as the earth from which they grow. Yellow is a fearful colour. It is associated with the deadly disease, yellow fever. The young unopened buds of the palm fronds have a mixture of this colour and green and are used for important communication functions. Blue as well as its different shades especially the dark blue is one of the principal colours used in dyeing native cloth. Since it is close to black it is often treated as belonging to that family and is consequently used for similar purposes as black.

Extra-Mundane Communication

This is a mode of communication, real or imaginary, believed to take place between the living and the dead, or between the living and supernatural or supreme being. This mode of communication is significant from the point of view that there is no society where it does not exist in its different forms. On the surface it usually seems unidirectional but participants at religious crusades, *idiong* consultation prayer sessions, rituals and other religious and pseudo-spiritual activities know there is often a form of feedback which may come through intra-personal processes, physical revelations or magical, other-worldly verbalizations. Thus it is a multi-directional or multi-dimensional mode which has become pervasive in all societies. It is possible to evoke such intensity of spiritual feeling through a spiritual transmigration of the participants to another world. Such a performance may convey the elements of a cultural celebration, dedication and consecration as is often witnessed in marriage and funeral rites or at (the pouring of) libation. Obituary, in memoriam notices, and tombstone messages are the graphic forms of this mode of communication. Among some of the other well known forms of this mode are incantation, chant, ritual, prayer, sacrifice, invocation, libation, conjuration, witchcraft, exorcism, vision and contemplation.

Institutional Communication

This involves the use of cultural or traditional institutions to communicate symbolically and as an extension of the extra-mundane mode of communication. The most important of such traditional institutions are marriage, chieftaincy, secret societies, shrines, masks and masquerades. There is a lot of communication in the enactment of activities of each of the institutions. For example, the marriage institution and all ceremonies pertaining to traditional weddings are a combination of a secular celebration, spirituality akin to christian ritual and a cultural communication of norms and mores of the society or group.

In addition, as Okita (1982: 9) pointed out,

> The importance of shrines as a vehicle for transmitting a people's cultural heritage does not only emanate from the fact that shrines are an embodiment of socio-religious ideas that give meaning and sustain life in a traditional society. The objects kept in shrines, the worship or festival that takes place there, are all symbolic acts that sometimes deal with the tradition or origin of a people, the rather inexplicable natural forces or phenomenon that must be personified: or certain norms or laws considered necessary for the sustenance of society all of which are invariably connected with the life cycle.

The ways in which all the above modes operate in the society show that they are all interwoven with all other human activities such that sometimes it is difficult to distinguish between the role of the traditional communicator as a practitioner and as an individual acting in his or her private capacity in society. And Pye (1963) observed that:

> Traditional communication processes . . . tended in general to be closely wedded to social and political processes that the very act of receiving and transmitting messages called for some display of agreement and acceptance.

This is true when a case like that of Chief Anangabo Effiong, of Mbarakom, is considered. He was a former traditional communicator who was so close to the political system and power base that he today is the village head of Mbarakom where he has served as media practitioner. The communicator, the medium, and channel in traditional society almost appear indistinguishable in their functions in society. The communicator can be seen as channel as well as medium; while the channel can be the medium as well as the message; so too can the medium be.

Thus, from the evidence on traditional modes of communication that have been identified above it is obvious that the various techniques and functions of communication in traditional society were not in themselves exclusive. Most media display the capacity for multi-social functions and the choice of each medium may be determined by the nature of the communication matter and the ability of the medium to get the message across to the audience in good time. The role of each medium of communication is thus circumscribed by the institutional and social control exercised on traditional communicators who themselves seem to be deprived of their individualism or that they have to subsume it within the context of a collective communal fiat; and also in the performance of their duty have to acquiesce to the demands of their job which has no place for self-interest or competitive adventurism as witnessed in modern media practice.

It is a historical fact that the various traditional approaches as described by the scheme set out in Table 3.1, to communication were as effective in reaching the small populations of the rural societies as our modern media are able to reach the teeming populations of the present age in urban centres. Today not all of the media mentioned are as effective in their coverage and function as they were in the past. It is for this reason that there seems to be much sense in suggestions for a multi-media approach to our communication problems in Nigeria and in Africa in general.

In spite of this fact the traditional media and channels of communication operate independently and also complement one another in the communication network that exists in rural society. Nevertheless, it is clear that the presence of modern media in all sections of society — rural and urban — has not in any way diminished the role of traditional rulers as gatekeepers in the information dissemination process, mediating between the larger machinery of the state and the gentle but determined murmurs of traditional media, since a large majority of the people do not possess the means to buy newspapers, radio, television or books and magazines. So many factors are at play in both the rural and urban areas which are not within the ambit of this study. But if there is anything this researcher has learned in this study, it is that he has got so much to learn.

TABLE 3.1

Summary of Traditional Modes and Instruments of Communication

		Idiophones:	Wooden drum, Woodblock, Ritual rattle, Bell, metal gong, akankang, ekere, xylophone, hand shakers, pot drum.
A	Instrumental	Membranophones:	Skin drum
		Aerophones:	Whistle, deer horn, ivory tusk, reed pipe.
		Symbolography:	Decorated bamboo rind, nsibidi, tatoo, chalk marks.
B	Demonstrative	Music:	Songs, choral and entertainment music.
		Signal:	Canon shots, gun shots, whistle call, camp fire.
C	Iconographic	Objectified:	Charcoal, white dove, kolanut, cowtail, white clay, egg, feather, calabash beads, limb bones, drinking gourds, calabash, flag.
		Floral	Young unopened palm frond, okono tree nsei, nyama, mimosa, plaintain stems.
D	Extra-Mundane	Icantatory:	ritual, libation, vision.
		Graphic:	Obituary, *in memoriam* notices.
E	Visual	Colour:	White cloth, red cloth
		Appearance:	dressing, hairstyle.
F	Institutional	Social:	Marriage, chieftaincy.
		Spiritual:	Shrine, masquerade.

Notes

1. Idiong is the most powerful secret and sacred cult among the Ibibio.
2. Interview with Edidem Ekpe Obong Atakpa, Paramount Ruler of Itu.

PART II

EXPLORING APPLICATIONS

Chapter 4

THE USE OF INDIGENOUS ENTERTAINMENT FORMS IN DEVELOPMENT COMMUNICATION IN GHANA

K. N. Bame

Introduction

Nowadays, some development strategists in Africa and elsewhere in the developing world tend to pay some attention to indigenous cultures. Western and urban-oriented development strategists who viewed indigenous culture as a "bulwark of conservatism" which hinders rather than promotes development and therefore must be ignored in development programmes are being ignored. Instead, the critics call for modifications to these cultures to support development and change.

Cultural variables are now perceived as very important in any attempt to generate behavioural change among people and therefore they should receive due attention in any development efforts. Thus Colletta (1980: 17) rightly stressed that

> a culture-based development strategy enables new knowledge and skills, and attitudes to be introduced within the framework of existing knowledge, cultural patterns, institutions, values and human resources. Indeed, the indigenous culture is the fabric within which development can be woven.

This recognition of the importance of cultural variables in development has generated world-wide interest in the use of traditional media, especially, the performing arts in the promotion of development.

Development is centred around people and it is now acknowledged that, in the language of former President Nyerere "People cannot be developed they can only develop themselves." In other words, peoples' participation and self-reliance are vital ingredients in development. This fact, in African context, makes apparent the catalytic potential of folk media such as drama, theatre, folk plays, oral tradition, in brief, folk entertainments or popular theatre, in promoting such self-reliant development. Being media wch address local interests and concerns in the language and idioms which the people are familiar with and understand, they are appropriate communication channels for the people. Any development-oriented ideas disseminated through them are more likely to involve and motivate a larger number of people to participate in the development process. Moreover, research has revealed that the success of development requires popular participation and "any communication media that are popular tend to be more effective in stimulating such participation than ones that are not" (Ingle 1972: 5). The research is often guided by theoretical viewpoints some of which are examined next.

Theoretical and Socio-Cultural Bases for the Use of Folk Media

A brief examination of the theoretical and socio-cultural bases which underlie the use of popular theatre in development in Ghana and elsewhere in Africa reveals a couple of empirically-based generalizations derived from communication studies which contribute to the explanation of or underpin, the use of folk media as appropriate channels for disseminating development ideas. They are that:

1. In order to convey a message with effect, one has to use the language, symbols and styles familiar to the audience (Ingle 1972: 30).

2. Attitude change is more persistent overtime if the receiver actively participates in, rather than passively receives the communication (Zimbardo and Ebbesen 1969: 23)

These findings on the indigenous media have convinced scholars like Ingle (1972) of the advantages for using traditional media in promoting development in largely rural and illiterate population as we have in Ghana. In his view

> These traditional media are not merely a form of art expression but are a way of expressing knowledge in a manner which is acceptable and functional. The poorest man in the land has access to his culture, expressed either in the story, poem, play, song, proverb, custom, ritual, rites or a variety of other ways characteristic of folk culture (Ingle 1972: 29).

Socio-Cultural Advantages of Traditional Media

These theoretical ideas are reflected in the following perceptible advantages which folk or traditional media have over modern mass media in promoting development in places like rural Ghana. The advantages obviously constitute reasons why media have received widespread use in development campaigns. Compared with modern mass media, the folk media are more familiar and closer to the ordinary people at the grassroots level and this fact would seem to make them more effective channels through which the ordinary folk can be presented with new and development ideas such as modern family planning. Being personal forms of entertainment as well as channels of communication, the folk media, such as traditional drama, story-telling and folk singing, are an effective part of the way of life of the people and thus provide fruitful means of disseminating ideas to them.

Again, being grassroots entertainment media they cover primary and intimate social groups and any messages they carry reach such groups and therefore reach the well-established communication network of any community.

Traditional modes of communication deal with the values and beliefs of the people and this would seem to make them useful means through which social engineers can bring about behavioral changes in people such as adopting family planning practices. This is because peoples' values and beliefs play vital role in their acceptance or rejection of such innovation as modern family planning.

Unlike modern mass media programmes which are usually produced for large and diverse audiences, the folk media can use local dialects to disseminate ideas in a most intimate and down-to-earth way at the village level in the rural areas.

Popular Theatre and Development in Ghana

It is such apparent advantages inherent in traditional media that have inspired numerous experiments in which scholars have attempted to encode development messages in such media especially, folk or indigenous drama to induce attitude and behavioral changes and thereby facilitate development in such areas as literacy, child care, nutrition, co-operation, sanitation and family planning practices. Two of the studies conducted in Ghana have been documented.

Chronologically, the first of them is what Pickering terms "Village Drama," (Pickering 1957a) an effective development-promoting folk drama used by Mass Education teams in Ghana in the early fifties. Pickering rightly perceives village drama as "the most truly Ghanaian audio-visual aid, depending as it does upon a nationwide aptitude and liking for drama and by its intimate relation to local custom and tradition" (Pickering 1957b: 178)

Being realists who were operating within the tradition of folk plays, the actors (members of the mass education teams) did not look for a sophisticated stage of a city theatre type which would not be available in rural Ghana any way. Rather they used as their stage any open space between two trees, a village square or a "clearing in the crowd." Equally simple were the stories dramatized. The stories which centred around the general themes of literacy, child care, co-operation and village health and sanitation, like the stories of concert part plays to be examined below, were unwritten. Members of the teams discussed the story or plot line and action of the play, allocated parts to the actors who were free to 'do their own thing' or place personal interpretation on their roles during the performances.

A popular play which the mass education teams often used to teach co-operation was entitled *Unity in Strength*. The play begins with an old man teaching his quarrelsome children a lesson about strength in unity. He uses a familiar household utensil, the broom, as a teaching aid. He unties the binding and gives each child a strand instructing him to break it. Each of them does so "with contemptuous ease." The old man reassembles the remaining broom, ties it with the binding and passes it round the children with the same instruction as before. Much to their chagrin none of them is able to break it. The father draws the moral, harmony is restored and the play ends with a song stressing the usefulness of co-operation and the strength in unity.

This short play of about twenty minutes duration gave ample room for amusement and buffoonery, especially in the quarrelling of the opening scene. It also provided opportunity to teach one of Africans' cherished values — traditional respect for the old, in this case, the father whose wisdom has been practically demonstrated. Gerontocratic as this might seem, the old man's brief homily in which he reassuringly drew the moral drove the lesson home not only to the audience but also to the actors themselves, mass education teams, whose desire for united effort in their work was sometimes obstructed by disruptive disputes among them.

Two of the stories which the teams dramatized in their literacy campaign are telling. In one of them, an eligible young bachelor is asked by a father to choose one of his two daughters. One is an illiterate but very beautiful and the other is plain but literate. The young man chooses the plain rather than the beautiful one because she is literate. The beautiful one realizing her handicap promptly enrols in a literacy class.

The second story is about a chief and his elders who act wrongly in a sale of land because they are illiterates. They sell a 20-acre plot of stool land to a prospector and seal a document on the deal by putting their thumb-prints on it. They subsequently discover to their disappointment and embarrassment that they have sold him 200-acres which a large diamond concern is trying to acquire. They send the case to court and justice prevails.

The story ends with the chief and his elders, all diligently attending mass education classes in order to become literate.

Our third theme of the village drama deals with rate-collection. It is described in detail if only because it provides a very good illustration of how a series of related messages can be encoded in a folk play to generate development. The story is about rate-collection in a very difficult area where the people did not want to pay their rates or levies, where the people complained that their rates were higher than people in other areas and that they did not have pipe-borne water supply as people in a nearby village had. Thus the first part of the play deals with the importance of rate payment.

The hero of the play is one Kofi Basake, a no-nonsense self-appointed leader of his community. The play opens with Kofi loudly condemning the local council in his area to friends in his house. A rate-collector arrives at the scene to collect Kofi's rate. In indignation Basake throws the intruder out of his house and in pursuing him into the street Kofi falls into a gutter injuring his leg in the process. This was enough for him to make an enormous fuss. He is very frightened at the sight of his own blood (red ink) and so he entreats his friends to take him to the clinic where he is attended by a spotlessly uniformed nurse who assures Kofi that there is no cause for alarm. Basake is moved to ask who provided the clinic. He is told it is the local authority which provided it and the nurse's salary, uniform as well as the equipment of the clinic. Kofi asks the nurse whether she herself has to pay rates; she produces her receipt, previously obtained from the council's treasury.

The play ends with Kofi Basake who has learnt his lesson the hard way, wrapped in bandages making peace with the rate-collector and dutifully paying his rate (Pickering, 1975a: 181)

As it happened, in the same area, there prevailed an unfounded belief that people who collected and disbursed rates could do as they wished with the money. It was thus necessary to keep the target population informed of three points. First, rate-collectors in the district bonded themselves for a large sum of money on appointment against any misappropriation of funds on their part. Secondly, the council meetings, during which decisions on the disbursement of revenues and other matters were taken, were open to the public; and thirdly, some council expenditure unsanctioned by the Ministry of Local Government could be and were recovered from the individual councillors themselves.

In view of the fact that the play about Kofi Basake was well received and enjoyed by its audiences, the plot was adapted to include the three points. The first point was made in the opening scene where Kofi quarrels with the rate-collector and the second was made in a scene in which a member of his ward persuades Kofi Basake to attend a council meeting and the third was made in a scene showing the actual council meeting: The three points were again stressed by the converted Kofi Basake to his friends. Toward the end of the play the converted Kofi Basake returns to his village and scolds his former co-belligerent for their ignorance and shares with them his newly-acquired wisdom.

This use of village drama was so successful and effective that according to Pickering (1975b), statistical records showed considerable increase in rates paid in 1953 and 1954 in the areas concerned.

A number of reasons account for the success. The first of them is the self-identification of members of the audience with the central character. Kofi Basake's problem depicted in the first scene, was the problem of members of the audience, his reactions to it were also largely theirs, and as the play unfolded and diverse influences were brought to bear on him the audience too were influenced, and at least for while they shared his final change of heart. Thus the Kofi Basakes among the audience were affected and they changed for good. Another reason is the usefulness of using drama for teaching an indirect lesson through a third person on the stage. This enabled the team to deal with a contentious subject which would easily arouse feelings in a real situation. In the play the councillor could tell Kofi Basake that he was ignorant and reactionary and that his influence was a serious hindrance to the development of the village, whereas if he had so censured a village Kofi Basake in a real situation that would have been disastrous for the objectives of the whole campaign.

Concert Party Plays

The second documented study on the use of folk media in family life education was conducted by the present writer. In the study, an attempt was made to tailor the message of modern family planning to suit the rural and largely illiterate population of Ghana by resorting to traditional mode of communication in presenting the message to them. The study was funded by the Population Dynamics Programme of University of Ghana.

Before the study, the channels of communication which had been used then in the communication of family planning in Ghana had been radio and television broadcasts, mobile cinema shows, newspaper articles and advertisements and handouts written in English and local languages and individual personal contacts by family planning workers. Traditional modes of communication such as folk drama, story-telling, folk singing and dancing activities had not been used.

The Experimental Study

In our experimental study, we chose two tradition media of communication — Concert Party Plays and town or village group discussions to communicate family planning. The principal objective of this pilot study was to carry out a field experiment in the communication of family planning through traditional media indicated above and to compare those media with the modern mass media so far employed for the same purpose.

Reasons for Using the Two Folk Media

In pursuance of the objective above, we set out to study the most productive folk media or a combination of media which could be used effectively in presenting family planning ideas to the large population of illiterate and rural people of Ghana. Village group discussion, a time-honoured medium through which rural Ghanaians communicate ideas, presented itself as a promising medium to use.

In addition, it is an empirically verified fact that discussions more than lectures encourage the involvement of the audience in communication transactions. We therefore reasoned that group discussion was more likely to lead to a high degree of personal involvement by the participants and make them become committed to the ideas and decisions arrived at during the discussions. This commitment in turn would lead to a change of attitude and behaviour. As regards the *concert-party* plays, we chose them because they effectively combine visual and oral effects in driving home their message. Moreover, they are also familiar to and highly popular among Ghanaians, especially among the rural and illiterate people, who are the target population.

Another reason for using the comic or *concert-party* plays comes from an empirical study previously conducted by the writer. As a part of a study (Bame 1972) of the diffusion and social functions of "comic plays" in Ghana, the writer conducted a survey on comic play-goers' reactions to the plays and the plays' influence on their attitude and behaviour. One of the hypotheses tested was that "the incidents which Ghanaian comic play-goers see in the plays on the stage influence and guide their daily lives."

A sample of approximately 1,000 play-goers was interviewed in all the nine regions of Ghana. One question framed to elicit information to verify the hypothesis was "Do the morals or advice which the comedians give at the end of their plays help you in daily life? If yes, how?"

The information gathered supported the hypothesis and clearly showed that the comic plays radiate conditioning influence on the thinking and attitudes of Ghanaian play-goers and that this influence, to some extent, guides and regulates their daily lives and behaviour.

Yet another reason for using the comic plays is that they can be made deliberately educational, and thus adding family planning to their content would make them an effective and more powerful medium for disseminating the new ideas.

Methodology of the Study

The study was oriented to diffusion of innovation theory and the field techniques used in collecting the data were observation and interview using interview schedules. The respondents were a probability sample selected by means of systematic sampling procedure, using as a population universe, listed houses in six Ghanaian communities (Adabraka, Greater Accra Region; Tsito, Volta Region; Essarkyir, Central Region; Effiduase and Kuronum, Ashanti Region; Badu, Brong-Ahafo Region) with the married people in the sampled houses as respondents. The study was carried out in two phases (i) the field experiment and (ii) the survey. The first part of the study covering a period of four months involved the exposure of the would-be respondents in three of the sex communities to the experimental treatments — the performance of a concert party play based on family planning ideas in the form of story summarized below and group discussion (also in three of the selected communities) led in each case by two trained members of the staff of the planned parenthood Association of Ghana in the regions concerned.

In the first month of the first phase, the concert party play was performed once in each of the communities and the family planning discussion group in each community began its fortnightly and guided discussions a couple of days after the performance of the play in the community.

Sampling Procedures and Sample

Although married individuals aged between fifteen and forty-five years were the respondents and units of analysis in the study, listed house of the six communities were the sampling units. From the listed houses systematic probability samples of houses were drawn by means of a table of random numbers, and the married people in the sampled houses in each community constituted the respondents.

The total number of houses listed in all the six communities was 2,185. One hundred and forty, a little over six percent of them, were selected by means of proportional stratified probability sampling the criteria of stratification being region and type of community (that is, urban or rural). Table 4.1 on the following page gives the samples for the rural and urban communities, which were chosen separately using different sampling fractions.

TABLE 4.1

Proportional Stratified Samples of Houses in the Six Selected Communities and Samples of Married Persons who were Covered by the Study

Community	Listed Houses	Sampling Fraction	Proportionately Allocated and Adjusted	Married Persons Who Participated in the Study		
			Samples	Male	Female	Total
Tsito	300	$\frac{300}{768}$	50	108	211	319
Essarkyir	103	$\frac{103}{768}$	22	44	128	172
Kuronum	96	$\frac{96}{768}$	21	65	87	152
Badu	269	$\frac{269}{768}$	45	78	186	264
URBAN SAMPLES						
Adabraka	675	$\frac{675}{1,417}$	60	181	297	478
Efiduase	742	$\frac{742}{1,417}$	73	203	395	598
	2,185		271	679	1,304	1,983

Research Instruments

A ten-page interview schedule containing thirty-eight items was the principal instrument used for this study. A shorter four-page version of the schedule was designed for the focused interview. The following are items 12 and 14 which were crucial for the study and are relevant for the analysis in this article.

12. Which *two* of the following media have helped you most to know about the benefits of family planning? (Put "1" against the first and "2" against the second.)

   1. Radio
   2. Newspaper
   3. Kakaiku's concert-party play
   4. Television
   5. Group discussion with family-planning workers
   6. Mobile cinema shows
   7. Family planning handouts
   8. Individual discussion with a family-planning worker
   9. Other, specify

14. If you have adopted family-planning practices or you plan to do so in the near future, which *two* of the following media have helped you most in your decision to adopt family-planning practice? (Put "1" against the first, etc.) (The same media items as in item 12 above were used.)

A Summary of the Story

The play which was performed by Kakaiku's Concert Party depicted the contrasting life styles of well-planned, well-organized and well-disciplined family on one hand and that of disorganized, impoverished unplanned family on the other. The first part of the play depicted a couple who had planned their family, had two sons and a daughter and had given them good education. One is a medical doctor, the other a lawyer and the girl a professionally qualified nurse. All three live abroad in Britain and they have come home to visit their parents. The life style of this family is enviable: they wear good clothes, eat good meals and punctuate their merry-making with intelligent discussions about the welfare of their town and family. Before the children return to Britain, they decide to build a better house for their parents. They each contribute his or her quota toward the construction of the house. Just before their departure, each of them including even the youngest, the nurse, gives their parents some pocket money. Delighted and almost overwhelmed by the kind gestures of their children, the thrilled couple say *bon voyage* to their happy children.

In contrast, the second part of the play depicted another couple who have not planned their family according to their means. They have ill-fed and ill-clothed eight children with the mother expecting another baby. The scenes for this family begin with one in which a guardian of one of the eight children, apprenticed to learn fitting work reporting to the father, Kofi Ataapim, that his son has made a secondary school girl pregnant and the father of the pregnant girl has been threatening to take the case to court. Kofi Ataapim should accompany him, the guardian, to go and settle the case. Money is needed for this but Ataapim does not have it and so he tells his friend to go back and expect him in a few days' time, hoping he will then have secured a loan from a friend for the settlement of the case. Just then, another son of Ataapim apprenticed to learn carpentry comes to tell his father that he has mastered his trade and his master wants Ataapim to pay the apprenticeship fees.

This money too will have to be borrowed by Ataapim from a friend. While Ataapim is informing his son, the carpenter apprentice, about his financial problems, the only educated child of his, a daughter, attending a teacher training college, arrives with yet another unwelcome news that she too has been sent home to collect her fees before she will be allowed to sit for her teacher's certificate examination. Ataapim now inundated with demands for money does not know what to do.

As if these do not constitute enough and disturbing financial problems for Ataapim, his educated daughter who has gone to see her expectant mother at the hospital returns with the news that Ataapim has yet another mouth to feed: her mother has had a new baby and the hospital authorities want him to go and pay the hospital bills before his wife and newly-born baby will be allowed to go home.

Ataapim now fumes with anger for the situation he has created for himself by not planning his family and bluntly refuses to go and pay the hospital bills, saying that his wife and the new baby could remain in the hospital; 'when the doctor becomes fed up with their presence he will send them home'.

Meanwhile, Ataapim's friend visits him and Ataapim recounts his financial problems to him. The friend advises Ataapim to stop producing any more children. He could obtain help in that respect from the family planning clinic. Ataapim invites his friend to share his kenkey meal with him. While they eat, all the eight children come in turns peeping and expecting to get some of the kenkey to eat. Both Ataapim and his guest friend are embarrassed and the friend stops eating and Ataapim also stops eating. As if by some pre-arranged signal or order, all the eight children, most of them in thread-bare clothes, storm the eating place struggling for the leftovers of the kenkey and literally fighting to obtain some of the food to eat.

The embarrassed friend of Ataapim leaves and another friend comes to enquire about what is happening amidst such shouts and noise over food. He too advises Ataapim to stop producing any more children because all his problems are the consequences of not planning his family.

The play ends with a song by all the actors asking all people to plan their families because having too many children leads to the children being ill-clothed, ill-fed and ill-educated.

Data Analysis

The main statistical measure we used to test whether or not there was non-chance relationship between variables was the coefficient of correlation called Yules Q^1. When necessary we resorted to the use of percentage differences to Supplement Q.

In line with the conventions for describing the values of Q we considered the relationship between any two variables as statistically significant when their Q values was $\pm.10$ or higher, indicating in each case, of course, the strength of the relationship, as Davis suggests. In other words, when the value of Q for any of the predicted relationship in the hypotheses was $\pm.10$ or higher, we concluded that it is significantly different from zero, meaning there is a no-chance relationship between the variables, thus indicating the confirmation of the hypothesis.

Verification of Hypothesis

The following is one of the four hypotheses verified in the study.

Hypothesis One: Urban respondents will name modern media as the source of family-planning information more often than rural respondents. Conversely, rural respondents will name traditional media as the source of their family-planning information more often than urban respondents.

Efiduase in Ashanti and Essarkyir in the Central Region are the urban and rural experimental communities, respectively, which were exposed to both the concert-party plays and the town or village group discussion treatments of the study. Thus, we used data from these two communities to verify the hypothesis.

The questions used to elicit information in the main survey to verify this hypothesis are questionnaire items 12 and 14, given above. (The data were cross-tabulated with type of community or place of living either urban or rural.) Of the various media items listed in question 12, Kakaiku's concert-party play and individual and group discussions with family-planning workers constituted traditional media. Thus, the number of times they were each mentioned in a community as either the first or second medium which had helped the respondents most to know about the benefits of family planning were summed up to obtain the score for the traditional media. Similarly, radio, newspapers, television, and the remaining items in question 12 constituted modern media; and the number of times they were each mentioned in a community as either the first or second medium which had helped the respondents most to know about the benefits of family planning were summed up to form the score for the modern media.

Through this process, we obtained the scores for modern and traditional media for the urban community Efiduase and the rural community Essarkyir shown in Table 4.2 to verify the hypothesis as shown below:

TABLE 4.2

Place of Living by Media that Influenced Decision to Accept Family Planning

Media	Urban Community (Efiduase)	Rural Community (Essarkyir)	Total
Modern Media	175	26	201
Traditional Media	205	86	291
Total	380	112	492 = N

770 = Total   278 = Inapplicable

$$Q = \frac{(175 \times 86) - (26 \times 205)}{(175 \times 86) + (26 \times 205)}$$

$$= .477$$

This value of Q (+.48) shows that there is a significant but moderate correlation between living in an urban community and naming more often modern media as the media that have helped one most in one's decision to adopt modern family planning practices. Conversely, there is a significant but moderate positive correlation between living in a rural community and naming more often traditional media as the media that have helped one most in one's decision to adopt modern family-planning practices. The hypothesis is thus supported.

Summary of Findings

The findings which emerged from our four confirmed hypotheses are that (1) literate respondents as well as urban respondents tended to indicate more often modern mass media items as the source of their family planning information; conversely, illiterate respondents and respondents living in rural communities tended to indicate more often traditional or folk media items as their source of family planning information; (2) literate respondents and respondents who perceived modern family planning as being good or of advantage to them tended to have adopted it more than illiterate respondents and respondents who did not perceive it as good or of advantage to them.

Further analysis of other aspects of the data yielded the following results: Indicating the media that had helped them most to become committed to family planning, the respondents gave the following ranking; first, radio; second, concert party play; third, cinema; fourth, group discussions with family planning workers; fifth, individual discussion with family planning workers, sixth, discussion with a spouse, a friend or relative; seventh, television; and eighth, family planning handouts. However, respondents who actually saw the concert party play ranked it first.

With respect to their effectiveness in conveying family planning message to the people studied, the two folk media items: the concert party play and the village or group discussions compared favourably well with the items of modern mass media; they were surpassed only by radio and mobile cinema respectively. Thus, whereas the concert party play proved to be specially suitable for disseminating family planning ideas to rural dwellers, the group discussion seemed to be suitable for both urban and rural dwellers but even slightly favoured by urban dwellers. Further research is needed here to get the explanation. However, the general pattern is that urban respondents tended to rank high modern mass media items whereas rural respondents tended to rank traditional media items high with respect to their influence on their attitude change in connection with family planning.

The respondents in the experimental communities, that is, those who received the experimental treatment showed more favourable attitude and seemed to be more committed to family planning than those in the control communities, that is, those who did not receive any treatment.

Taking a lead from these findings and available literature, the present author, at the time of writing in early 1993, had secured funding from African Development Foundation in Washington D.C. to use folk drama to promote the use of Oral Rehydration Therapy in the treatment of diarrhoea in rural Ghana. The main objective of the study is to carry out an applied research based on the findings of previous studies by using folk drama and discussion groups to communicate oral rehydration therapy in selected Ghanaian communities and assess their effectiveness.

Three of the specific questions which the study seeks to answer are (i) As compared with other media items (such as radio, newspapers, television, mobile cinema shows and ORT handouts), is folk drama more or less effective in communicating ORT messages in Ghana? (ii) What are the comparative effectiveness of folk drama and group discussions in the communication of ORT messages? (iii) Is folk drama more or less effective in communicating ORT messages as it is in communicating family planning messages?

Undocumented Use of Folk Entertainment Forms

In the Promotion of Development in Ghana
Although the two studies of the use of folk drama in the facilitation of development examined above are the only documented ones in Ghana which have come to the author's notice, there are several groups in Ghana currently using folk drama or entertainment forms in similar manner. The following are among the undocumented lot in Ghana now:

Ghana Congress on Evangelization

Women's Ministry
This group has branches all over Ghana which use drama to educate adult women on family life and problems. Two examples of their plays were performed at the University of Science and Technology, Kumasi and a village in the Central Region of Ghana.

Church Drama Groups
These exist in a number of churches in Ghana, example being the Pentecostal Church. They perform plays on family life education for the benefit of their congregation and non-members.

The World Vision International

Women's Groups

This has women's development section of the organization co-ordinated by a lady by the name of Mrs. Alice Yirenkyi. She uses folk drama as "starter" in women development programmes. According to the co-ordinator by "starter" she means an initial activity which she uses, as it were, to brainstorm the women to enable them to identify their family life problems and later gear their efforts towards solving them. Such a starter enabled a women's group at Asesewase in the Eastern Region of Ghana to learn the benefits of co-operation among themselves in their daily life activities. A similar drama has enabled a Labadi Women's group to identify various trades or vocations in which they are currently engaged and earning their living. In all these dramas, the performers or actresses are selected from among the target women group itself. The co-ordinator merely serves as a facilitator for the group.

Social Welfare and Community Development

As to be expected, the Department of Social Welfare and Community Development whose mass education teams were the first to employ village drama in the fifties in their development activities still use folk plays in the nineties for the same purpose. They use folk drama to arouse and sustain social awareness of their target populations and mobilize them for local development.

Atwia Story-Telling Group

The last but not the least group whose folk entertainment art form can be adopted for development is Atwia story-telling group. Atwia is a farming village in the Central Region of Ghana whose inhabitants have formed a story-telling group headed by the female chief of the village.

About the middle of April in 1988, at the invitation of the institute of African Studies, the group came to present an exhilarating performance of their art at Legon. During an interview at the end of the performance, their leader explained to the present writer the way they use their art form to promote development. Being a popular indigenous art form, story-telling pulls crowd. So they seize the opportunity in the course of their performances to focus on important messages on development to the large audiences.

It is apparent that this art-form, like other folk entertainment art-forms, has a great potential for effective promotion of development which, so far, does not seem to be exploited. Messages on family planning and other areas of development can be encoded in the stories themselves. Messages thus encoded will reach ordinary people at the grassroots level and motivate them to get involved in development.

Owing to the fact that folk entertainment forms have been found to be an effective tool for self-reliant development, they have been and are being used in the promotion of development in other African Countries such as Nigeria, Botswana and several others.

A Critical Review

These various ways in which serious and innovative attempts have been made to facilitate development by spreading development ideas through folk entertainment forms are refreshing. The lead they provide holds promise for current and future development activities in third world countries, like Ghana, which are predominantly rural. However, lest an impression be created that such uses of folk drama or entertainment forms have no demerits, a critical look at them at this point seems expedient.

Although they have been able to achieve their objectives by effectively disseminating information on development to their target populations, the studies examined above, including the undocumented uses of folk entertainments in development generally adopt what may be termed a "one-shot" approach. The visiting theatre groups involved utilize folk entertainment forms or popular theatre during a period of time to disseminate ideas concerning family life education or other facets of development among their target populations, change their attitudes and behaviour relating to particular issues and end the process there. They fail to involve members of their target populations in the creation and performances of the folk drama or entertainment forms. Thus they do not leave their target populations with any performance skills or legacy.

Obviously, they need to go beyond this one-shot approach and involve their potential audiences in the creation and performances of their drama or entertainment forms so that they do not just remain members of the audience, but become actors.

In that way, the local actors can build on their newly-acquired skills and in all probability form their own theatre groups. In that position, they will have the required skills and be ready to use the groups to mobilize their people for development as and when necessary. In other words, outside development-oriented theatre groups should leave behind in the localites where they work some structures for future self-reliant development activities.

One other criticism often levelled against the use of folk plays in mobilizing people for development is that such plays which disseminate Government-sponsored development information do not provide real dialogue in which local people participate in the form of bottom-up communication. Instead, such media are simply used in the same top-bottom fashion characteristic of modern mass media. The centrally determined information which they disseminate, which does not give the rural folks any opportunity to critically analyze their content tends to make them passive and dependent on the ruling or dominant class who use the popular theatre to reinforce their privileged position. The people's own indigenous theatre is in this way appropriated to spread ideas which may not be in their interest. The net result is that the people's theatre is used "in their own domestication" — it becomes a very effective means of socializing them to accept their situation without complaint or reflection. As Ross Kidd puts it, "By turning the folk performers into mercenary propagandists, folk media experts undermine their credibility and destroy people's culture" (Kidd in Ross Kidd and Nat Colletta 1980).

Indeed, folk drama can either be used as a tool for mobilizing people for development such as family life education or it can be co-opted by interest groups such as the ruling class in a country to foster their own interests, for example for maintaining themselves in power. This is a real danger. In order to avoid such a situation, popular theatre experts should work with the people at the grassroot level, identify their problems and use the medium to harness their energies to solve those problems. Hence, the caveat to popular theatre experts is that they should guard against using folk drama to cultivate or reinforce the 'culture of silence'. Rather they should aim at using it to promote a genuine development of the people.

Yet another criticism against folk plays is that the plays emphasizes self-reliance and help members of a village or community to come together and solve their problems themselves, as a result they tend to deal with problems which can be solved locally and leave untouched more serious problems which affect them and people in other regions, for example high cost of living. However, it should be noted that some popular theatre organizers especially in South America, have broken this chain of parochialism and have attempted to address wider problems, which were perceived as political and which thus worried political authorities who resorted to repressive measures to silence the organizers. Oft-quoted African example of such organizers is a Kenyan playwright Ngugi Wa Thiongo who suffered imprisonment for his popular theatre activities.

Implications For Policy and Practice

The main findings of the studies, summarized above convey a number of practical implications for development action programmes for Ghana and other third world countries. But one general implication which seems immediately apparent is the need for varied and multi-source approach in the presentation of development messages to suit the different areas of Ghana as well as different sections of the target populations. This implies that the whole array of both modern mass and folk media items analyzed in the studies and others which may not have been mentioned must be employed. The multi-source approach will not only take account of the rural-urban and literate-illiterate variabilities with respect to people's attitude toward acceptance of different development projects which have been empirically substantiated in our analysis, but also it will take into account of a possible variability between persons living in the same community.

The practical implication deducible from the general ranking of the various media items in the family planning study is that the radio, the concert party play (or folk drama in general) the mobile cinema and group discussions stand out as the four most effective media for communicating family planning in Ghana. Thus if people in charge of family planning or other development programmes for any reasons have to select only four media items for their communication and motivation campaigns then the findings of the study suggest that the four indicated above are to be recommended. On the other hand, if bearing in mind the rural-urban dichotomy they decide to divide the four media channels between rural and urban areas, then the two most effective media channel for urban communities are the radio and group discussions in that order whereas the two most effective channels for rural communities are concert party plays or folk drama and the mobile cinema also in that order.

The village drama study and indeed other studies conducted in the area elsewhere, all convey similar messages for policy and practice. They all confirm that folk drama or folk entertainments, in general, have a great potential in the field of community development in Ghana and Africa generally.

In fact, it is apparent from our survey of the role of popular theatre in development communication in Ghana that there is no better and effective approach to facilitate development among rural Ghanaians than this culture-based development strategy. Our analysis has shown that popular theatre activities constitute an effective medium for communicating development ideas to Ghanaians; such activities hold promise for future development in third world countries. Their potential must be further harnessed through innovative adoption to generate self-reliant development in Ghana and other developing countries.

NOTES

1. A couple of the characteristics of Q which are advantages to its use are these: (1) Unlike Gamma, which may be used only if all or at least one of the investigator's variables are of ordinal or higher, "Q may be used on any variable regardless of the level of measurement" (Davis 1971, p.75). (2) Again, unlike Gamma and other measures, Q is margin free, that is, it does not vary with the size of the marginal proportions.

2. It should be noted here that a respondent will only name a medium as a source of his information if he has ever been exposed to it. Thus, that fact is assumed in this hypothesis.

3. In line with Ghana Population Census and other conventional definitions we regard Efiduase in Ashanti as an urban community because it has a population of over 5,000 (the actual population in 1970 was 6,967) and other services like a post office, a commercial bank, a hospital, or health centre, and so on. Similarly, we regard Essarkyir in the Central Region as a rural community because its population is *less than* 5,000 (the actual population in 1970 was 1,779) and it does not have the services just indicated above.

Chapter 5

THEATRE FOR DEVELOPMENT AND THE EMPOWERMENT OF DEVELOPMENT SUPPORT COMMUNICATION IN AFRICA*

Kees P. Epskamp

It has been many years since development workers in Africa began experimenting with the techniques of development supporting theatre. "Theatre for development," as a special form of popular theatre within the field of adult education, is meant to be a community focused problem-solving cultural intervention strategy in which the process of creating a play is educationally as valuable as the product: the performance.

Media such as theatre and other performing arts might be used to release the wealth of knowledge that exists in a country, and in the process give people an active part in development support. This approach requires the use of media at the local level so that the contents of programmes can be based primarily on real-life situations of people.

In the traditional education of children in the rural areas of Africa, music, dance and drama were taught. From their infancy, knowledge and skills in these arts were transferred during work, leisure, on festive occasions or during games. Instruction became more deliberate in the initiation camps, where the main teaching tools were imitation and repetition. Besides this rather classical type of instruction, each young person being initiated had his or her own individual supervisor. If someone showed particular talent in one of the performing arts, s/he was put under the tutelage of a master.

During the colonial period, the education of children was primarily in the hands of the missions. As in Europe, the church made use of "educational drama." On the one hand this built upon the skills in the performing arts that the children had already acquired at home; at the same time it offered a "respectable" alternative to the "heathen" dance dramas and ritual masquerades that were native to the children's communities. It did no good to forbid dance dramas; the dance groups simply "went underground." So why not turn sin into virtue? In the 1950's, for example, the Holy Cross

* This chapter is developed from a combination of three chapters of "Learning by Performing Arts; from indigenous to endogenous cultural development" written by Kees P. Epskamp and published in 1992 by the Centre for the Study of Education in Developing Countries (CESO), The Hague (Paperback no. 16).

mission post in Pondoland made use of dance drama to teach the story of Saul and David. The fight with Goliath was especially suited to incorporating local war dances (Taylor 1950: 298–300).

But most useful of all for purposes of preaching the gospel and educating were the narrative songs and sketches. In South Sudan, converted Dinka minstrels were put to work preaching the word of God. They made up their own new words for the refrains of old songs, and the audience sang along. Experiments in using sketches were tried out mainly in schools and urban neighbourhoods. For the pupils it was a good way of being introduced to the written word. In Uganda parables were acted out. The teacher told the parable and the pupils improvised a sketch around it. Discussion followed and changes were made. Costumes and props came into the picture only at the final rehearsal.

In the 1950's there were also performances at the mission schools outside regular school hours. Works performed were often those by European authors: Gilbert and Sullivan operattas, 18th century comedies and the classics. Texts were memorised, European theatrical conventions were used, and local black people played the roles of white pirates, shepherds, and baronesses. The teacher followed the script word-for-word and there was no room for improvisation.

The present-day educational system in Africa is still strongly marked by the colonial past. Most African states acquired their independence in the historic period of the 1960s. However, this by no means eliminated all traces of European influence as French, English or Portuguese continued to be used in the schools; and pupils were expected to sound as British or French as possible. The criteria for theatre were drawn from the literary traditions and ideas of Europe, hence students tended to be scornful of their own traditional culture, becoming alienated from their own society by the time they finished their schooling.

In the 1970s people in Africa began to question what the place of music, dance and drama should be in regular school curricula as drama lessons tended to be more incidental than a structural part of education.

Now, there are efforts to make drama more than just an extracurricular activity in African schools. The people who design teaching materials are gradually appreciating the value of traditional educative games and drama.

Botswana is where it all Started

Within the African context, the pioneering role of "theatre-for-development" projects in Botswana during the seventies cannot be ignored. The initiative of the Botswana case and others charged the objectives of several University theatre companies during the seventies. The assumption of being a travelling circus to provide the country with culture gave way to making plays with the rural population on the accepted realities in the country.

Because the Botswana initiatives of the sixties were generally perceived to have been successful, they were incorporated into all kinds of other projects. One of them was a theatre for development project especially designed for the Sarwa (the Bushmen). The Sarwa are hunters and food gatherers who used to inhabit a large part of Africa but who were driven back more and more by the surrounding stock-breeders and farmers. Forced or voluntarily, many have given up their former existence and are now employed by the Tswana — the main population of Botswana — or the whites as herdsmen or farmhands. Few still live the original way of life.

In the seventies, research showed that many of them would have liked to have a plot of their own which they could cultivate independently and where they could keep cattle without being exploited any longer by the more wealthy farmers. The government responded by apportioning plots to them in the so-called state lands and provided these with wells.

However, it was a big step for the Sarwa to change after so many years from a state of dependence into a completely independent life. In order to bridge this gap, the government organized a series of workshops for each group of new settlers, prior to their departure for their new community.

According to Kidd and Byram (1977: 27), the Sarwa remained rather passive. Then the idea arose to apply the principle of participatory drama, which proved to be very successful. In Sarwa culture there are rather a lot of performing-elements which lend themselves admirably to such an approach. Many Sarwa people have a flair to imitate situations and behavioural patterns. The Sarwa can tell each other stories for hours on end or perform dance dramas in which each Sarwa, man, woman or child is allowed to participate.

Participation of the local population was stimulated during the entire process: from preparing, executing and evaluating the discussion to follow-up actions carried out. In 1974 and 1975 the evaluation of these theatre projects was primarily aimed at the organisation of the festivals. In 1976 and 1977 its emphasis moved on to the execution and results of these educative activities. A comparison of attendance figures showed that the festivals were far more successful than the barren meetings of, for instance, the extension officers of the health-care sector. The participation in the preparation and execution as well as the actions actually carried out by the people after the festival were indicative of the the effectiveness of the performing arts as a motivating stimulus for mobilization.

Drama Teacher Training at University Level in Nigeria

All Nigerian states have their own Council for Arts and Culture. In general these councils are meant to preserve the local culture. The main activities of such a council consist primarily of organizing traditional performances or parades for visitors and tourists; promoting or preserving the production of traditional artefacts; and booking a local singing and dancing group to welcome dignitaries at the airport. In the beginning of the eighties, the Benue Arts Council gradually became aware of the need for a progressive popular culture alongside an operative traditional popular culture and an elitist westernized culture. The Council's new strategy was not only to serve the traditional folklore and culture, but also to sustain the prevailing cultural expressions of the rural population, the Tiv.

For this reason the Benue State Council for Arts and Culture organized a course for its own staff members and for a number of social and cultural workers of the Ministry of Social Welfare, Youth, Sports and Culture.

Participants were trained in the use of theatre for educative purposes and for "awareness raising." Apart from these officials, students of social academies, health care workers and regional radio officials took part. The course was led by "resource people" teaching at Ahmadu Bello University in Zaria. A constant interaction with the local Tiv communities was stimulated. For the students this workshop was meant as a "training of trainers;" for the local people as "instruction and entertainment."

The bilingual workshop was meant for adult educators. Two techniques were used: improvisation and elocution. In the workshop the following construction was used: collection of data in the area (action research), analysis, improvisations, performances and, finally, discussions. The results of a workshop depended mainly on the motivation of the individual participants. There was a clear discrepancy between participants sent by their employer and those who took part on a voluntary basis. Whatever intentions the participants had, they all received a certificate of attendance at the end of the course.

In general it was not easy to present theatre in the Hausa states of northern Nigeria, such as Zaria, Kaduna and Benue. The population is made up largely of Islamic Hausas with small enclaves of other peoples. A devout Muslim sees theatre as exhibitionism. And under no circumstances should women appear in public. In many situations, there was clear hostility.

In one village, stones were thrown at the students for play-acting in public. In another, they were allowed to perform, but the chief forbade the use of drums. Again, all open-air performances had to be scheduled for 3.30 pm after work and afternoon prayers, and not afterwards.

The attitude of the drama students towards making theatre for and with rural communities was often not without contradictions. Some stridently preached about the class struggle, thinking that this was required or enabled them get good grades, or simply because it was fashionable.

The course leaders tried to make both the students and the audience see local problems as more than merely topics for anecdotes. The object was to force them to make a deeper analysis.

Malawi: University Theatre and Censorship

Kerr (1981: 47–48) distinguished three trends in the popular theatre in Malawi. First there was the elite literary theatre in English which, for want of television, was rather popular with a small group of well-educated people. Secondly there was the, as he called it, "populistic" theatre. This theatre is tuned to the average Malawi population but openly follows the government policy or makes itself subservient to national campaigns. Finally, Kerr (1981) mentioned a genuine folk theatre, which apparently was apolitical and used all kinds of authentic elements of the theatre, such as singing, dance, music and dialogue.

Kerr (1981: 47–48), in his discussion of theatre, believes the political circumstances of the country one is dealing with should be taken into consideration. Malawi provides us with an example of how a growing post colonial literary tradition for years conflicted with the National Board of Censorship. The country gained its independence in 1964 and, until recently, had a president as head of state who prefers to use a one-party system, the Malawi Congress Party (MCP), of which the president himself is the leader. All the country's media, including plays and theatre performances, are under direct control of a Board of Censorship.

Also, according to the Law of Censored Supervision over Entertainment, a play can be a tragedy, comedy, acting, opera, farce, revue, variety, comic interlude, melodrama, pantomime, dialogue, prologue, epilogue, poetry, lectures, visual designs and song texts. The law prohibited in these media any public display of, or reference to sexual, religious or political affairs. With this, it was almost impossible for writers' and actors' workshops, literary student magazines and theatre groups to turn out any products. In 1972, only four out of eight suggested plays were passed for performance.

Of these, one was immediately suppressed after its first showing. This situation lasted until 1978.

It was obvious that the reintroduction of this law mainly aimed at shutting up those media which had a considerable audience. The literary activities at the university were isolated from the general public. The university, established in Blantyre since 1964, and its allied colleges, mainly thought of theatre activities as extracurricular. The only college which officially had a Drama Department was Chancellor College. As a sub-section of the Department of English, it regularly turned out theatre productions, such as *Everyman* or Brecht's *Caucasian Chalk Circle*. Although the language used on stage was English, these plays toured the country. To make the text less inaccessible to the rural population, key lines were vaguely rendered in the local languages such as Chewa.

In 1973, the University was literally isolated from the general audience. Colleges were fused and moved to Zomba. An open-air theatre was built, which became the base of the new Chancellor College Travelling Theatre. However, increasing cultural suppression and censorship between 1974 and 1976 made it impossible to choose from the repertory developed from various world popular theatre workshops or to work on performances based on improvisation. The only way to continue productions was by the use of allegories and symbols which only insiders might interpret as hidden political criticism. Censorship led to ingenuity. And it was not surprising that the resistance against this suppression stimulated the use of allegories, usually derived from indigenous metaphors and endogenous resistance during the colonial past.

The growing number of elaborate studies of the Chewa and Man'ganja rituals and the mythic sources of the Nyau cults were a stimulus here. To the government officials, this literary development was a welcome phenomenon at a time of the Africanisation of history. Not until the mid-seventies were indigenous performing arts acknowledged as a source of inspiration for literary development in Malawi. The number of performances based on the indigenous culture, symbols and allegories increased remarkably.

Towards the end of the seventies the pressure of censorship in Malawi declined. Production of plays based on improvisations performed in local languages and denouncing social abuse actually began in 1981. In that year, *Eviction* was produced. It was about a teacher who had been evicted by the Ministry of Housing. Thereafter, the Office of the President and the Cabinet (OPC) invited the university theatre to play in Mbalachanda, a newly opened rural centre in a remote part of the Mzimba district.

The invitation would only be accepted on condition that plays took the form of workshops, to be performed for and by local people. Consent was given and together with the rural population, a number of plays with themes such as literacy and sanitary problems, landed property and landless labourers, as well as activities of the Agricultural Information Service were developed and staged. The performances were shown in several surrounding villages. However, the discussions with the audience afterwards were somewhat problematic because they were usually dominated by a number of Malawi Congress Party loyalists.

Two problems thus immediately emerged: how to reach the local people without the interference of the village head and how to clarify that the students had not come to solve the problems of the people. The students only wanted highlight by acting out local problems in order to facilitate their discussion. Whether the rural populations were fundamentally helped by the workshops and performances was unclear. What was clear was that a practical follow-up in setting up a programme of activities to actually support the local people was beyond the capacity of the University as a training college.

National and Community Theatre in Namibia

The National Theatre of Namibia (NTN) is based in Windhoek, the nation's capital. Large-scale and international theatre productions such as the ANC's *Amandla* are performed there. The NTN is an independent business that has financial support from the Ministry of Education and Culture. According to Zeeman (1990: 38), the goal of this cultural institute is to support indigenous performing artists to do productions. The National Theatre also houses its own group and, finally, it accommodates so called "high culture" productions.

However, the relevance of the NTN has never been certain. Does it serve the interests of a modest cultural crowd in Windhoek or those of the entire country? It also happens that the latter is not really within the financial and logistic capabilities of the NTN. That is why the rural communities sum up the relevance of the NTN in Windhoek as being "next to nothing." In spite of this, the NTN works hard to become a "flagship resource centre" for all Namibians who are involved in the performing arts.

Because 98 percent of the Namibian population is Christian, the first community performances were passion plays organized by the church. Later, school performances followed; then performances that originated from extramural activities were initiated by enthusiastic teachers. This was how the Rehoboth Community Theatre Group began.

In Namibia, each community celebration is graced by music, dancing and singing. In the northern region, in particular, a strong dance-drama tradition remains. Gradually it appeared that the most creative potential in Rehoboth could be found in the poorest part of the community, in Block "E." From way back, Namibian cultures such as the Khoi, the Herero, the Nama, the Himba and Ovambo, have known a rich narrative tradition. Block "E" was, after all, the melting pot of the most diverse ethnic backgrounds from all parts of the country. It is no coincidence, then, that the Rehoboth Community Theatre Group was born in that block. Whereas in other parts of the *baster* community drama was not regarded a serious activity for adults, for the black society, performing music or a little mime was as natural as their other daily activities.

The closure of mines near Rehoboth has led to an increase in structural unemployment. It has created a *baster* society with social problems. Men now leave their villages to work somewhere else. Women stay behind with the children. When the mother, heading a one-parent family, also has to provide an income, the children remain at home unsupervised and do not complete school. In some neighbourhoods, gangs of youths, often under the influence of alcohol, indulge in petty crime. There is a stabbing almost each week. Among the girls, teenage pregnancy and prostitution are common.

There is also a racial problem in the form of a feeling of superiority by *basters* over black ethnic groups. This can be explained historically. In the past, *basters* often needed to defend their land and cattle against nocturnal robbery by surrounding black neighbours with a warlike culture who were also cattle breeders. The *basters* became so effective in these frequent armed conflicts that the German colonial regime engaged them in private armies to protect their goods and chattels from the black population. This has not helped the *basters'* popularity among their black neighbours.

Following the whites, the *basters* also engaged the local villagers to do inferior, domestic and menial work. This is how domestic personnel in Rehoboth came to be put up in special accommodation known as Block "E." Block "E" residents are totally separated from the *baster* community. They live in abject poverty in slums made of cardboard and corrugated iron. Here too, there was mass unemployment among the black population. On the shrinking labour market, the *basters* and their neighbours compete for the few jobs that become available.

It was difficult enough to let the people from the other areas in the *baster* community talk let alone act their problems out. With the youth, though, it was not so difficult. A small theatre group had been formed at the secondary school of Rehoboth, and the members, under the inspiring leadership of their teacher, had produced a play about their home situation in the quarters. The play not only dealt with the usual youth problems of going out, dating and unwanted pregnancy, but also with problems such as unemployment and alcoholism which parents were also experiencing.

Adults in the areas did not particularly like having their problems exposed so publicly. From the start, it seemed the performance of a short play would not exactly lead to structural changes. This is where the didactic paradox of this kind of training seminar lies. For, the main objective is the training of cultural workers in techniques of problem-solving by way of a cultural intervention strategy. But, because of emphasis on the objective of raising awareness with limited time for action plans, the community often found itself out in the cold with unsolved "acted-out" social problems.

International Popular Theatre Alliance (IPTA) Meeting in Namibia

In 1983, some 70 drama teachers, development workers and actors from all over the world got together in Koitta, a small Hindu village 40 miles north of Dacca. The participants from Bangladesh were part-time actors and actresses of the Aranyak theatre group, Proshika animators and peasants running theatre groups in their own villages. The foreigners were all social and cultural workers from Africa, Asia, Canada and the Caribbean Islands. All participants were, in one or the other, practically involved with progressive and politically oriented popular theatre. Therefore, there was enough experience to share within the twelve days of the meeting.

In further pursuit of simplicity and relevance, this International Popular Theatre Dialogue (IPTD) was not opened by some Minister of Culture or Education, but by a landless labourer from one of the local small-scale and inexpensive awareness-raising theatre groups.

A concrete result of this meeting was the foundation of the International Popular Theatre Alliance (IPTA), an informal network of popular dramatists, whose office was to be moved every three years. In 1984, the Asian region offered to host the network for three years. The office was located at the Philippine Educational Theatre Association (PETA). The initiative received sympathy and modest support from the International Council of Adult Education (ICAE) based in Toronto, Canada.

The Philippine based secretariat initiated a permanent exchange of information and experience through the *International Popular Theatre Newsletter*, conferences and seminars. The Alliance aimed at promoting the exchange of teachers and animators working in the field of popular theatre in the various regions. On another front, it was realized that members who were involved with making activating theatre often run the risk of being imprisoned or banished on political grounds. Consequently the IPTA-secretariat players became engaged in presenting political petitions and declarations of solidarity.

For the second time, from August 14, 1991, Rehoboth, a small village 80 kilometres south of Windhoek, took its turn to host the International Popular Theatre Workshop. Similar to the meeting in Koitta (Bangladesh) in 1982, this event received technical assistance from the ICAE and was organized in close co-operation with the government of Namibia. The participants consisted of some 40 Namibians and 35 foreign guests.

The objectives of a workshop organized as part of the event were to:

1. train artists and community workers in participatory research and artistic skills.

2. give support to Namibian cultural workers to launch some national organizations.

3. review the work of popular theatre/popular education internationally.

4. promote creation of further regional and international linkages among popular theatre and popular education organizations.

5. strengthen the link between Namibian adult educators and their colleagues from within the region and outside.

6. hold a two-day cultural festival of representative groups from within the southern African region.

After two days of discussions, five days were spent working in the neighbourhoods of the rural town of Rehoboth, an ethnic enclave of coloureds in the middle of Namibia. The work consisted of participatory research into local social problem areas and cultural manifestations. This research, ultimately, resulted in short sketches produced together with the communities and was shown to a wider audience.

The event was rounded off with a party for all the participants which took the form of a theatre festival in Windhoek, where mainly groups from southern Africa performed. The ANC, for instance, brought a performance of *Amandla*. Other performances were by the Zimbabwe Association for Community Theatres (ZACT), the Zambian National Theatre Arts Association (ZANTAA) and the Bricks Community Project from Namibia.

African Networking of Performing Artists

Jointly organized by the International Theatre Institute (ITI), and the International Amateur Theatre Association (IATA), with the participation of the government of Zimbabwe, a Consultative Conference on African Theatre was held under UNESCO, September 3–5, 1983. The University of Harare hosted a gathering of some thirty theatre workers, the majority from countries south of the Sahara.

To demonstrate their determination to give a voice to the African performing artists, and in order to defend more effectively their interests through collective action, the participants at the conference unanimously signed a charter. The charter gave birth to the Union of African Performing Artists (UAPA), a pan-African organization. Though created out of the initiative of theatre workers, it embraces artistes from other forms of performing arts such as music, dance, folklore and even cinema.

The Union aims to be interdisciplinary and non-political. It does not discriminate between the practitioner and the theoretician, the modern artist and the traditional artist; nor the amateur and the professional.

One of the objectives of the UAPA was the provision of training facilities for African performing artists, a training programme with a non-academic approach to training. The programme aimed at mastering both quality in performance and development supporting issues in which the trainee learned technical and communicative skills. These were enabling skills geared toward improving one's performance, work management and the overall development of the community, culture and nation.

For this reason the opening of an African Centre for the Training of Performing Artists (ACTPA) was included on the UAPA programme of action with the active support of the then vice president of the Union, Daniel Labonne. The creation of this training centre received priority on the agenda.[1] Part of the strategy of ACTPA was to link local training needs with international policy issues. UNESCO liked this strategy. However, it provided a lot of sympathy but little financial support.

Obtaining the sympathy of the UNESCO International Fund for the Promotion of Culture, a programme for the development of the performing arts in Africa was initiated. ACTPA fell between the framework of the objectives of the World Decade for Cultural Development (WDCD) on the one hand, and the priority given by UNESCO to development in Africa during the Medium Term Plan (1990–1995), on the other. The latter programme is based on south-south relationships because of the almost general absence of horizontal exchanges among African countries, particularly in the performing arts. African artists have greater opportunities to be known outside the continent on the occasion of international festivals and meetings or concerts organized by Western promoters, than by fellow Africans in other countries of the continent.

With this moral support from UNESCO it was up to UAPA to formulate a plan for a training programme, find a location and a financial credibility to be sponsored by international donor agencies. Hence, there was a need for a private and independent intermediary organisation, a consultancy agency. For this reason Daniel Labonne started a modest bureau called African Theatre Exchange (ATEX), based in London, close to the north European donor agencies. There the elaborated plans for ACTPA were born and workshops were programmed of which the African Symposium Workshop (AFSYMWORK) in Mauritius (1988) is the most well-known example.[2]

During this workshop the plans for the pan-African training centre for performing artists gained ground. The long term objective of such a training centre should have been the promotion of the African performer, allowing him or her to play a more responsible and effective role in the society. On an even larger scale the training programme aimed at restoring the educative and cultural role of African performing arts in the future African society.

The government of Zimbabwe felt very strongly about housing the African Centre for the Training of Performing Artists in the Castle, located in Bulawayo, Zimbabwe. This attractive building is self-contained with facilities for offices, workshops, recreation rooms, board and lodging for the students and a performing area. The Ministry of Education and Culture of the government of Zimbabwe was willing to take full responsibility for housing the training centre on the explicit condition that it would be financed by donor agencies from the north. This should not have been a problem after ACTPA/ATEX already staged two training workshops, both ending with a well-received performance: the dance drama *Footprints* (1990), partly based upon Footprints about Bantustan by Tafataona Mahoso, and *Lucy & Me* (1991).

According to the Zimbabwe Ministry of Education and Culture, these two productions brought together several eminent African theatre trainers and performers from fifteen African countries and cultural backgrounds. In the production of *Lucy & me*, theory and practice brought together anthropologists and performing artistes. Lucy, the small hominid of bones which was discovered in Hadar (Ethiopia 1974) described as the mother of all ancestors is estimated by academicians to be 3.5 million years old. Thus, in African terms, Lucy is the female creature entering mythology.

Although these two performances were received with a lot of sympathy, the Zimbabwe Ministry of Education and Culture at the end of 1991 could not guarantee the follow-up of the Bulawayo based ACTPA because international donor agencies refused to commit themselves to supporting the project for the duration of at least three years.[3]

This was unfortunate because, within the southern African region, Zimbabwe is one of the most active countries exploring the effectiveness of theatre for development within national campaigning as well as small-scale local development communication supporting projects. For example, the Kenyan refugee Ngugi wa Mirii, theatre coordinator for the Zimbabwe Foundation for Education with Production (ZIMFEP), initiated a countrywide network of community based theatre companies which are strongly interested in theatre-for-development techniques.

Development Supporting Theatre in Times of AIDS

According to McIvor (1990: 29) the observation that: "Give an audience a lecture and they will listen. Give an audience some theatre and they will participate . . . has been established among community theatre groups throughout the globe for quite some time but it is only in the last few years that ministries and agencies in Zimbabwe, seeking to promote better health, literacy, and social awareness, have come to this realization.

These community theatre groups in Zimbabwe try to make the general audience aware of the problem of how to handle physically disabled persons within the community as well as how to handle other outcasts within society. These groups are also active in health care information campaigns which aim at preventing venereal diseases and AIDS.

Gortzak (1992) and others have noted the presence and quick spread of AIDS in Africa. Part of the problem of controlling the disease is attributed to the taboo on talking about sex. Within the limits of normal social conventions in many parts of Africa those men who are not having sex regularly are not normal; they are not "real men." "Talking about sex" in Africa is something people (men as well as women) are not used to. "Sex is penetration. Hugging is suspicious, something typically Western."[4]

In the view of Gortzak (1992: 7), what culture and history have to do with this is that in the rural areas of Africa, ideas about potency and fertility are issues villagers do not talk about easily in spite of the importance of these issues in their worldview and belief system. Traditionally, lifelong monogamy has never been a strong cultural issue in Africa, although in the past only those in power (the rich and well-to-do) could afford polygamy and, thus, polygamy was practised by a very small elite.

In spite of the problem with public discussion of sex, people tend to be prepared to think about these subjects differently in the face of the AIDS epidemic. It still turns out to be very hard though to persuade men as well as women to use contraceptives, especially condoms. For prostitutes who have to survive on sex advice to use condoms may not be taken seriously.

During colonial rule, the practice of forced labour in the mines, as well as forced migration due to the worsening situation of the labour market, meant that men left their household for quite some time and would live with other women in their new location. After a while, these men would return usually with sexually transmitted diseases contracted at the mines.

These days, because of the explosive AIDS situation, several foreign non-governmental organizations (NGOs) and sponsors are co-operating with African governments to change this situation by creating some awareness about AIDS. Some of these agencies in their extension work use theatre and other performing arts to at least inform people.

In Zimbabwe ZIMFEP theatre coordinator, Ngugi wa Mirii, started an AIDS awareness theatre in conjunction with the Zimbabwe Association of Community Theatre (ZACT). Their play on AIDS was called *Manyanya* ("It's too much"). It is a theatrical production which combines drama, humour, music, song and dance. The play revolves around the sexual behaviour of the central character: a factory manager. It unfolds a pattern of how he contracted sexually transmitted diseases (STD) and eventually AIDS. The drama focuses on how his wife, a teacher, and many other people contracted either STD or AIDS through their sexual partners. The play thus sought to promote monogamous relationships.

The play also dramatically reveals the struggle to look for a cure; the reaction of members of the main character's family and workmates after they learnt that he has AIDS. ZACT was able to set the play in both urban and rural Zimbabwe. That way it cut across the social strata of society, touching on the lives of the young, old, rich and poor alike. The impact of the play is strengthened by the use of songs and dances deeply rooted in the various cultures of Zimbabwe. However, in southern African cultures including the Xhosa, local health workers are not encouraged to discuss sexuality in the open.

That is why the theatre, *Puppets Against AIDS*, is trying to bring some relief in this area of preventive health care. The programme is an initiative of the African Research and Educational Puppetry Programme (AREPP) founded in 1987. AREPP is a non-profit educational trust based in Johannesburg. Its main objective is to provide educational puppet theatre and training workshops concerning issues related to the environment and the physical well-being of the communities in the southern African region.

Unlike the Xhosa health workers, the puppet programme has the license to speak openly in public places, and to reach the masses without being offensive. Two versions of the performance have been offered so far. The first version featured two-metre high puppets, with a human animator hidden inside the puppet. Currently, smaller glove puppets are used. This makes the show more portable because it requires only a single puppeteer, a narrator and a technician, thereby decreasing costs.

To eliminate any possible racial associations with the disease, the puppets are grey in colour. The story is related in the local language by the narrator, while music, either live or recorded, provides a background. Performances are followed by a condom demonstration, question-and-answer sessions and the distribution of condoms and illustrated "AIDS talk" comic brochures in the local language.

More to the north, in Zambia, various groups have worked in the field of preventive health care, especially AIDS prevention. In the western province a Dutch-sponsored project was elaborately described by Determeyer (1992). In this area the Zambian government and various missionary organisations in collaboration with a network of eleven hospitals, ninety three health care centres and two hundred community health workers units started experimenting to use community theatre in disseminating AIDS prevention information.

Summary

This chapter is intended to contribute both to the theory and the practice of theatre as an educative medium to be used in processes of social change and development in Africa. It hopes to achieve this by placing popular theatre and theatre for development in the context of history and local culture. "Development co-operation" as a post-war phenomenon became current after most of the present countries in Africa became independent in the fifties and sixties. This implies that "theatre for development" has a history of some four decades.

The discussion here mainly focused on an overview of initiatives which were undertaken in several African countries, to introduce theatre for development to the rural communities in the 1970s. Most of these initiatives were launched by a small group of foreign employees attached to the departments of English or Drama, which were sometimes also called the department of "Dance, Music and Drama" or the department of "Performing Arts." The staff members of these various universities knew each other or had met each other at a later stage of their career. Hence the influence on one another's approach was inevitable.

In theatre for development, both theatre-makers and development workers want to convey defined messages with relevance, and designed, for specific audiences.

During the last two decades, governmental and multilateral organizations have taken it for granted that theatre is a relatively inexpensive educative means which uses the language of the people. Theatre seems to avoid the problem of illiteracy, is part of the local culture, and is a form of entertainment. Traditional theatre has been used rashly in the framework of information campaigns about health care, hygiene, agriculture, birth control, and political lobbying.

These ideas of the use of the performing arts as popular media in non-formal education have been developed mainly by NGOs and institutions for adult education. The departments of adult education carried out experiments in rural development, integrating the performing arts in non-formal education activities, such as literacy programmes, community development and adult education.

A problem which was repeated often during the experiments in the field of rural development and adult education was finding a balance between the social and the artistic criteria that are required for this kind of theatre. Despite this, the emotional impact of a play which is based on an existing social relevance, was often thought to be more important than the artistic criteria.

The plays are still often initiated and performed by people from a community different from target local communities. Among the first to realize that the use of theatre in this way would not have any educative impact were the cultural and social workers, the animators in the field. Engaged in awareness-training about unequal social conditions, they were less concerned with telling people how to increase their crops by using fertilizers as with the problem of gaining access to the people through whatever channels. To these workers, not every form of theatre could be effective in their work. Why not? Because some theatrical forms were embedded in a grid of socio-cultural values, or were directly associated with previous national information campaigns involving, for example, regard to family planning.

For example, in the eyes of a devout Muslim, theatre is considered as mere exhibitionism. Thus, under no circumstances are women allowed to watch or even mingle with the male-dominated audience. In this context it is not easy to present theatre in the Hausa states of Northern Nigeria.

Using local theatrical forms of expression, theatre groups in developing countries seem to be achieving an impact unachievable by Western-oriented theatrical concepts. On several occasions traditional theatre has been used for information campaigns concerned with health care, hygiene, agriculture, birth control, and political canvassing. It needs to be noted, though, that any arbitrarily chosen traditional theatre form will do as a medium whether it is used to inform people about the use and results of fertilizers, or make them aware of the fact that the use of fertilizer is going to make them dependent on banks and multinationals.

In this context the preceding sections on Botswana, Nigeria and Malawi show how vulnerable these forms of theatre might be. Apart from taking the time to convince the audience that the actors are not hired as political fawners, there is always the chance that by lack of strong political parties or unions to support this kind of theatre, the actors and community members would receive little protection from potential reprisals. Several socio-cultural constraints are to be noted. First, there are the authoritative bottlenecks either of a religious or of a political nature. For these local authorities the display of problems was one thing, action was another. That is why the performances are always critically observed by the local religious and secular leaders.

Although a lot happened in the world in the last fifteen years with regard to the organization of popular theatre, a clear philosophy and direction for the future appear to be lacking. It also seems as if the various African theatre groups and networks share a certain collective political orientation and commitment, although their practical activities and the development of concepts differ considerably. This not notwithstanding, Mwansa believes this diversity shows a certain extent of beauty.

NOTES

1. Those who formed the first board of UAPA were Debebe Eshetu from Ethiopia (as president), Daniel Labonne from Mauritius (as vice president and treasurer), Stephen Chifunyisé from Zimbabwe (as vice president), Hansel Eyoh from Cameroon (as secretary general), Sophia Lokko from Ghana (as member), Kalinguy Mwambay from Zaire (as member) and Penina Mlama from Tanzania (as member). Some of these people were directly involved in the ACTPA/ATEX training programme which

led to the staging of *Footprints*. Among them were Debebe Eshetu, Stephen Chifunyisé and Daniel Labonne. The other leading figures during this workshop were Amadu Maddy (Sierra Leone), Louis Akin (Ivory Coast) and Francis Nii-Yartey (Ghana).

2. ASYMWORK was organised in Mauritius from 15–29 October 1988.

3. In the past the ACTPA project received financial support from the national development co-operation agencies of all the four Scandinavian countries, as well as from those of Germany, the Netherlands and Canada. Rumour has it that this support stopped because of an ongoing fight between UAPA and ATEX. This discontent among the members and the board of UAPA and the director of ATEX caused international disinterest by donor agencies to allocate funds for an initial three year contract.

4. See an interview written by Hans van de Veen, entitled "Viola Mukasa uit Uganda" and published in *Onze Wereld*, 35(1992) 4: 34–36.

Chapter 6

FROM RITUAL TO THEATRE: VILLAGE BASED DRAMA FOR DEVELOPMENT IN AFRICA*

Louise Bourgault

Introduction

Clifford Geertz's *Ritual and Social Change: A Javanese Example*, written in the 1960s has become a classic work in symbolic anthropology. Geertz's account details the events surrounding the death of a Paidjan, young Javanese boy and the social disruption churned up as his uncle sought to have the appropriate death rituals performed. The case study analyses how and why the symbols bound up in the death ritual could no longer function well in parts of eastern Central Java by the 1950s and how the disintegration of a ritual form, the *slametan*, was related to social change occurring in the *kampong* of Modjukuto where the event took place (Geertz 1973)

Geertz's account serves as a kind of cautionary intellectual tale and reference to it is used to begin this paper for several reasons. First, it focuses on the importance of ritual for traditional people. Second, it suggests that in the enactment of ritual, the conflicts of a society are worked out or at least made palatable to the members of the group in the face of the broader issues of existence. Third, it provides a model of social action which takes into account, and gives equal weight to "logico-meaningful structures and causal-functional structures" in society and elaborates on the inherent differences in integration between these two systems of social functioning. Fourth, it locates ritual as a potential source of social change. And fifth, it suggests that societies experiencing rapid economic, social, or political change are likely to undergo difficulties as the cultural structures which provide meaning to people become detached from segments of the social grouping.

The purpose of this chapter is to introduce a discussion on the use of village based drama as a form of mass communication for development in Africa. The chapter attempts to accomplish a variety of tasks related to this theme. First, I will show why the use of theatre and other performative

* The editor gratefully acknowledges the cooperation of *The Journal of Development Communication* in allowing this chapter to be republished from the original article which appeared in its volume 2, number 2 edition, pp. 49–73.

genres have been neglected by development scholars. Second, I will examine the phenomena of ritual and attempt to show that it is closely related to theatre. I will argue that what can be achieved through ritual is obtainable in large measure through the use of theatre. Third, I will furnish a brief review of the literature on "theatre for development." Fourth, I will provide a case study examination of a village based drama production showing how many of its elements are related to ritual. And finally, I will comment on the findings in the case study in the light of ritual theory, cultural theory, and a definition of development.

The Neglect of Performative Genres in African Development

Since its birth in the late 1940s (Stevenson 1988), the field of development has been dominated by the social sciences: sociology, psychology, political science, and economics (Arnold, 1990). These fields all developed in the early 20th century and bear the marks of the modern paradigm with its scientific search for laws of development. Commonly, such laws were derived inductively from the examination of quantitatively based studies which used the individual as the unit of analysis. Through the years, these fields became increasingly technocratic, balkanizing themselves against the humanities (Arnold 1990) including the study of theatre arts.

The field of anthropology itself, such a seemingly obvious tool for development work, has also been neglected. Perhaps anthropology has been seen as too quirky, too esoteric, and rather too preoccupied with regressive exotica to be of much use to those who would see the exotic disappear through modernization (Hitchcock 1987).

Undoubtedly, the field of anthropology contributed as well to its own neglect in the domain of development studies, preferring to concentrate its analysis on the purest of traditional societies, those most likely to remain outside of the currents of modernization (Hymes 1972).

Anthropology, in any case, had by the mid-twentieth century fallen victim to the overdeterminism of structural and functional paradigms (Geertz 1973). But by the 1960s, the field of symbolic anthropology began to rediscover the niche carved out by the early historical anthropologists. Such scholars as Clifford Geertz, Gregory Bateson, Mary Douglas, and Victor Turner returned cultural systems, their subsystems, and performative genres, to their rightful place in the study of social action.[1]

The field of development communication has been, almost by definition, a *science* of development through the study of *mass communication*. Throughout the 1950s and 1960s, it focussed its efforts on determining the ways mass communication systems might bring about social change.[2] Since such systems tended to be located in the urban areas of Third World nations, those who studied the effectiveness of these systems tended to have an urban bias.

The diffusion of innovations perspective (Rogers 1983) of the 1960s represented an improvement over the earlier globalistic theories. It advocated the case study approach to development communication, thus returning some of the emphasis in development on smaller scale studies performed in the rural areas. Diffusion of innovations is concerned with the way mass communication currents ripple through a community. It has been sensitive to issues of message salience, adoption strategies, and interpersonal dynamics of innovative behavior. Still, its focus is rather too rationalist and rather too centered on the individual and its approach too unconcerned with group processes.

With the difficulties encountered in so many communication for development projects, new and improved methods emerged in the 1970s. Social marketing was derived as a technique useful in performing discrete communication tasks to solve specific social problems, such as the promotion of "oral rehydration therapy" or the advocacy of infant inoculation programmes. This approach and its offshoots, notes Hornik (1988: xii), focuses on "doing communication better." Borrowing heavily from American advertising practice, social marketing stresses the importance of audience research, project monitoring, and message pre-testing among message targets.

Though social marketing has had some successes, its focus is still far too concerned with individual targets. This has been evident in the reports of those who attempted to conduct target research in the service of social marketing projects (Obeng-Quaidoo 1985) or to teach the research techniques involved therein (Bourgault 1987). Indeed, social marketing along with its predecessors in development communication theory still ignores the altogether different *weltanschauung* of the oral world which characterizes many of the audiences of development communication, especially in Africa (Bourgault 1990).

A second genre of development communication projects also emerged in the 1970s. Hornik (1988: xii) has called these projects "interactive projects." These have been designed to more fully include recipients of the development in the planning. Many of these have been seen as "grassroots" projects, projects which might bypass urban elites in Third World societies, the latter often said to thwart development efforts, wittingly or unwittingly, through the diversion of funds, the creation of bureaucratic obstacles, or the advocacy of inappropriate development schemes. In many ways, the Liberian Rural Communications Network project of 1980–89 was based on the "interactive" model as the description of the experiment in village theatre described below will tend to indicate.[3]

Another reason for the neglect of performative genres in African development, it seems, has to do with their very nature. Much traditional lore in Africa is shrouded in secrecy.[4] Aspects of it arouse fear and dread (Turner 1982). Westerners working in development communication operate outside the currency of this knowledge. And many Africans, now working in the modern sector (which includes development projects) seem to live in a psychically bifurcated world, one in which the realities of village life are kept separate from the world of work in the modern professions. The reasons for this are bound up perhaps in their own deep ambivalence about rural life (Bourgault 1989). It may also be that like most westerners, they fail to appreciate altogether consciously the power village rites and rituals may have over them.

Ritual and the Theatre

The Importance of Ritual for Traditional Peoples
A ritual is patterned enactment (a set of acts/procedures) involving the performance during occasions not given over to daily and technical routine, of specified symbolic behaviors and the manipulation of sacred objects all of which are innately related to the society's beliefs about or its relationship with the cosmos. The performance of rituals is said to strengthen social solidarity by reminding performers of their place in the cosmos, their connection to universal forces, and their ties with one another. Thus, in traditional agrarian societies the important rituals are sacred ones.[5]

Sacred rituals include performative frameworks which help humankind to cope with life's awesome dramas: birth, social puberty, marriage, parenthood, social advancement, occupational advancement, and death (van Gennep 1960). They also include celebrations of the passing of seasons, as happens with calendrical rituals, healing ceremonies against sickness and infertility, and crisis rituals whose enactment may be used to reunite warring groups or factions, and droughts or pestilence. These rituals when taken together represent the cycle of life which is by definition sacred — bound up in the relationship between humans and their gods (Sieber 1987).

Rituals are particularly important in traditional agrarian oral societies because they along with myth are the dominant means through which the group is able to articulate its relationship with the universe (Langer 1957). They are also useful because they articulate in a dramatic way the social organization of a community. Through their enactment is demonstrated most vividly the place of everyone in a traditional community. As Chernoff (1979a: 161) notes, Africans like ritual because it is useful in "providing [them] a framework to help them know what is happening and to get into it." Social solidarity is strengthened through the enactment of ritual in traditional societies, because the social order present in the ritual form is made to be immutable. Phenomenologically speaking, these rituals present the world in the only way that it can be.

Through Enactment of Ritual, Conflicts Are Worked Out

The enactment of ritual must, however, occur in a place separated symbolically from ordinary technical, day to day existence. This separation from mundane existence occurs through "ritual framing." In Turner's (1982: 82) formulation, framing separates' "the indicative mood," or daily life, from "the subjunctive" or the period of ritual enactment. This framing is usually accomplished through "an act that clearly demarcates sacred space-time from mundane space time" (Turner & Turner 1988: 202; Turner 1982: 84) as when the prospective initiates in the Kpelle male initiation rite are physically taken from their villages into the Poro fence (Gay 1973, Bellman 1984).

The basic structure of a ritual is a tripartite one. Its three obligatory phases are the *pre-liminal*, the separation phase; the *liminal*, or marginal phase; and the *post-liminal* or aggregation phase. In the separation phase, the subject is somehow set off from the rest of society so that the ritual may be performed (Turner 1982).

During the liminal phase, the heart of the ritual enactment, actors live outside the boundaries, of the social world. They live at the margins. It is in this phase that ritual transformations of the actors occurs. This is accomplished in an atmosphere where the normal rules of everyday life have been suspended. Here usual everyday objects and practices have been made sacred by the ritual frame. The manipulation of these objects, their recombination and realignment, together with processual acts, play aspects using masks and costumes, creates the meaning shifts for the individual and for the group so that the transformative power of the ritual can occur (Turner 1982).

Turner (1982: 41) describes the atmosphere of the liminal phase of a boyhood rite of passage as

[Through] ordeals, myths, maskings, mumming, the presentation of sacred icons to novices, secret language, food, and behavioral taboos, create a wierd domain the seclusion camp in which the ordinary regularities of kinship, the residential setting, tribal law and custom are set aside, where the bizarre becomes the normal, and where through the loosening of connections between elements customarily bound together in certain combinations, the novices are induced to think [about their place in the cosmos].

In group rituals the marginal phase of a ritual engenders what Turner calls *communiatas*, strong feelings of 'solidarity in equality' engendered among ritual participants who have lost the social and psychological moorings of the indicative (and normally hierarchical) world and find themselves unleashed together in the often unsettling "subjunctive world of the ritual" (Turner 1969: 94–130). In the post-liminal phase, actors in the ritual are reunited with the larger society, the necessary transformations having occurred. The new initiate is welcomed as an adult; the formerly ill person is seen to be well.

In examining ritual, it is important to take note of their processual form. For this we must recall a portion of the definition provided for Turner (1982) and his precursors, "a patterned enactment." Rituals involve procedures set down by the genre. The ritual is like a road map marking the way to the accomplishment of certain goals: how to move from childhood to puberty; how to be transformed from a condition of barrenness to one of fertility; how to move the time forward from planting season to harvest time, and so on. In Turner's (1982: 80) formulation, the ritual steps are "*unidirectional and irreversible. The ritual is transformative because it is sequenced* [italics mine]."

Despite the patterned structure of ritual, however, there is also an inherent plasticity in its enactment. Indeed, Turner (1982: 81) says that rituals are not carved in stone because

[Most often,] invariant passages and episodes are interdigitated with variable passages, which, both at the verbal and non-verbal levels, improvization is not merely permitted, but required.

Thus, for Turner (1982: 85), this plasticity *contains the potentiality for cultural innovation,* as well as the means of effecting structural transformations within a relatively stable socio-cultural system [emphasis mine]. This is key to an appreciation of the role of ritual or ritual-like experiences in development.

Levels of Integration: Logico-Meaningful Structures and Causal-Functional Structures

Clifford Geertz (1973) would contend that the field of anthropology has been mired by structural-functional paradigms which deal inadequately with the phenomenon of social change. In such formulations ritual, myth, and a wider array of performing arts are viewed either as a mere reflection of the social systems, or as a source of social solidarity or individual balance. The cultural systems, the systems through which people weave "webs of significance," observes (Geertz 1973: 5), and hence build meaning into their world, are given second place to social systems by the structural-functionalists (Geertz 1973).

Geertz (1973: 144–145) argues that the cultural systems must be given equal weight with social systems in social analysis, that each must be regarded as "independently variable yet mutually interdependent factors;" noting further that both culture and social structure are different abstractions of the same phenomenon. Geertz (1973) also points out that cultural systems, logico-meaningful structures, possess a type of integration which is different in style from that of social systems. The differences between the two are that:

> Because the two modes of integration do not operate according to the same principles, because the particular form one of them takes does not directly imply the form of the other, there is an inherent incongruity and tension between the two, and a third element, the personality structure of the individual. In this tension there lives an energy which can be harnessed in the production of social change (Geertz 1973:145).

Rituals as Sources of Social Change

Since rituals function to align and realign the social world with the world of belief, their performance can be a powerful source of social change. In their enactment, strains are uncovered between the individual and the social world, between the individual and cultural beliefs of his/her community, or between the society's organization and its social practice, among other things. Because of the inherent plasticity of the ritual manipulations, some of these strains can be adjusted by the actors and new meanings can be formed in the process. And because of the rite's processual form, transformations of the individual or a social group necessarily occur in its enactment and are binding.

Ritual in Times of Stress

Societies undergoing rapid change are likely to "cling[s] noticeably to the symbols which guided [their] parents through life in rural society" (Geertz

1973: 165). This is no doubt why, in times of crisis, many Africans are said to supplicate their ancestors with special urgency and with a said increase in ritual activity (Douglas 1970: xix).

Theatre and Ritual: Similarities and Differences

Turner (1982) and Schechner (1985) show that ritual and theatre are very closely related. Turner (1982) maintains that both genres derive from what he calls social drama. Social dramas are conflicts which arise in the "indicative mode" of life. They constantly occur because individuals or groups suffer from competing loyalties. And these conflicts are never ending because they can never be completely resolved.

Among the Ndemu of Zambia, for example, a host of conflicts is built into the system which is organized according to the principle of virilocal marriage but matrilineal inheritance. In that system, divorce works in various ways to reassert the ultimate paramountcy of the matrilineal line (Turner 1982: 62). Indeed, Ndemu social life is fraught with conflict as members seek to marry, move to the home of the male partner, and yet pass on wealth through the female line. But modern Western marriage is also the source of social dramas as marrying couples cleave to their spouses at the expense of their birth families.

Turner (1982: 68) argues that all societies are rife with social drama; a formulation which implies that social dramas are a basic unit of social interaction, that every conversation is a potential conflict — a sort of drama waiting to happen as he writes:

> My observations convince me that it [social drama] is a spontaneous unit of social process, and a fact of everyone's experience in every human society.

Turner (1982: 76) describes further his conception of a social drama:

> . . . my social drama is agonistic, rife with problem and conflict, and this not merely because it assumes that sociological systems are never logical systems or harmonious *gestalten*, but are fraught with structural contradictions and norm conflicts.

Of this Geertz (1973) would say the modes of integration of the causal functional system and that of the logico-meaning system are different in form and style.

For Turner (1982: 78), rituals and other performative genres derive from these basic social dramas of which he writes:

> The social drama, then, I regard as the experimental matrix from which the many

genres of cultural performance, beginning with redressive ritual and juridical procedures, and eventually including oral and literary narrative, have been generated.

For Turner (1982), then, the source of ritual and the source of theatre, is the same, social conflict, or conflict between norms and social institutions or actors.

However, Schechner (1985: 22) believed the origins of both genres are kinetic, in movement rather than in social discourse; arguing that both theatre and ritual must have descended from the cave dances of our paleolithic ancestors. Both scholars nonetheless see the genres as intimately connected.

According to Turner (1982: 69), social dramas contain four phases: breach, crisis, redress, and either reintegration or recognition of schism. Redressive ritual operates in the third phase. It is a measure whose form is designed to repair an impending breech in the social fabric, or in an individual life. When it works, for example, when the gods are appeased through a patterned enactment of some sort, the actors of ritual are re-integrated. They have "gotten right with the gods" — a score between human participants or between the human and the divine has been settled. The society goes back to reasonably smooth functioning or the individual is cured. When the ritual fails, the society or the individual supplicant is still in crisis. Another ritual will be enacted until reintegration is believed to have occurred (Langer 1957:159). If the ritual fails for too long, a recognition of schism may obtain.

It should be recalled that rituals themselves have three phases: the pre-liminal, the liminal, and the post-liminal. As noted above, the heart of the ritual is the liminal phase, the phase where the indicative mood is suspended and where the transformations of the actor(s) occur, often in conjunction with other actors who share in the liminal phase and who together experience *communitas*, the social bonding derived from the shared experience of ritual participation.

It is during the liminal phase that *flow* is achieved. Turner (1982) derives his treatment of *flow* from Csikszentmihalyi (1974). For both these scholars, *flow* represents a phase in performative genres where the ego of the actor is lost through a merging with the experience. This can occur because the consciousness of the individual has been narrowed so that his/her complete focus is on the action of the enactment. Hence the actor "finds himself in control of his actions and of the environment." (Turner 1982: 57). Turner (1982: 57) adds that flow is coherent because the demands of the action are non-contradictory. The rules of the engagement are clear.

This control is also "autotelic," that is, "it seems to need no rewards outside itself."

Observing that the flow often occurs during trance-like states induced "through repetition and accumulation" Schechner (1985: 11), in his work with theatrical and ritual forms, devised all night dances to demonstrate the pattern of flow describing the experience, thus:

> Each time I've participated in this kind of dance I've had, and others too have had, a trance-like experience . . . where for varying periods the sense of me as an individual, the amount of time passing, the awareness of the environment I was in (outdoors in a field and inside a gymnasium to name two) were abolished. What was left were a vaguely recollectable sense of moving in a circle and the feel of other persons, the other bodies, to either side of me.

For Schechner(1985: 124) then, the *flow* is the portion of the ritual where the transformations occur.

Turner (1982: 121–122) says that for the modern individual, *flow* is taken up largely in liminoid genres. Artists become immersed in their creative activity, and make "art for art's sake" giving themselves over completely to the rules of the genre. He argues that aspects of that genre are reincorporated in a piecemeal fashion into everyday life by spectators and actors alike. This is in sharp contrast with aspects of ritual which are reincorporated as a whole into the lives of society members.

Schechner's (1985) discussion of reincorporation of performative genres is a bit more complete. Schechner (1985: 125) regards theatrical performances as *transportations* rather than *transformations* because professional actors do not emerge from a performance with a change in status.

> During the performance the performers are 'taken somewhere,' but at the end, often assisted by others, they are 'cooled down' and reenter ordinary life just about where they went in.

Schechner (1985: 127) further contends that because of the nature of acting, an activity in which different roles are assumed and are incorporated into the actor's being, "*a series of transformations can result in a transformation*" [italics added].

Schechner (1985: 145) argues that western *audiences* of conventional theatre (modern rather than post-modern) can only be *transported*, they cannot be transformed; mainly because these audiences have no role to play in the production so their involvement cannot extend beyond a "private sentimental empathy." Schechner (1985) contrasts this involvement level with that of some other modern, but non-Western theatre and with ancient Greek drama. In Japanese *Noh* and in the Indian mytho-religious enactments, the spectators actively *participate*. And in ancient Greek Drama, "the response of the judges to a particular performance determined who won the prize" (Schechner 1985: 148). These participatory theatrical genres, he maintains, *are transformative*. Clearly, his discussion suggests that the more involved the audience, the more likely is the possibility of its transformation. But surely, one observes, there is some transformative potential in empathy. For, Schechner (1985: 9) himself argues earlier in his treatment of flow that flow is the moment when the performance "takes off." when the audience becomes aware that something has happened. And here he suggests that in flow, not only do divisions between actors break down, but divisions between the audience and the actors can also occur.

Surely, then this question of transformation through involvement should be seen on a continuum. In such a model, the ritual has the most transformative power, modern western theatre the least, with "in between" genres offering up some interesting transformative possibilities.

Since the origins of ritual and theatre are one and the same, it seems likely that we will find other similarities between them. Turner (1982) treats this issue in a discussion of similarities and differences between liminal (ritual) and liminoid (art and play) genres. It is now possible to summarize the formulations of Turner and Schechner and to present them in a schematic diagram, one which clearly highlights the differences between liminal and liminoid genres (Table 6.1).

Through the combined work of Turner and Schechner it is now possible to present a bi-polar model, a summary of their formulations, in a schematic diagram which provides in clear relief the distinctions between ritual and theatre.

Liminal phenomena are typical of tribal and early agrarian societies. Here, the rituals are collective events. They involve the entire community. Participation in these liminal activities is mandatory as they are "centrally integrated into the total social process" (Turner 1982: 53).

Participation in liminal genres involves both work and play. Here it is reflective of life in the indicative mood, where the cycle of life is seen on work-play continuum. Liminal genres involve enjoyment and entertainment, but they are also deeply serious, often dreadful and terrifying.

Liminal phenomena use a common set of symbols, the meanings of which are shared by the entire group. Liminal phenomena, because they are collective, cement the social system.

TABLE 6.1

Schematic Diagram of Turner and Schechner bi-polar Model

LIMINAL	LIMINOID
Origin in Social Drama	Origin in Social Drama
Historical Peak: Tribal and Early Agrarian Societies	Age of Enlightenment
Involvement: Entire Community	Segment of Society at Fringes
Audience: Participant	Differentiated & Passive
Social Role: Integration of Total	Social Criticism of Community Segments of Society
Mode of Consciousness: Group	Individual
Combines Work and Play	Play only
Participation Required	Participation Optional
Involves Flow	Involves Flow
Is Deadly Serious	Is Entertainment/Diversion
Transformation Permanent	Transformation is Temporary, a Transportation

Liminal phenomena mute individuality, merging the identities of the actors with the action of the ritual, and hence with the cosmos and the society's belief about it. In so doing, they achieve flow, a kind of "self-propelled take-off," which results in transformation (Turner 1982: 53).

Liminoid genres include theatrical genres. They appear to have developed in conjunction with city states on their way to becoming empires (Ong 1977, Hogben 1969). They are in full development in modern industrial societies having come to full term during the Age of Enlightenment (Turner 1982: 53).

Liminoid activities are "quirky and idiosyncratic" because they represent the view of a single artist, "the glorification of the individual being the hallmark of industrialized society" (Turner 1982: 53–54). They occupy a position at the fringes of society rather than at its centre.

Participation in liminoid genres is optional, as these constitute a form of leisure which is separate from the serious work of society. Participation in liminoid genres is thus more transitory and far less serious than in the liminal. And liminoid phenomena are commodities to be enjoyed during leisure time, which is functionally segmented from work (Turner 1982). Schechner (1985) would add here that in modern theatre, audiences are functionally separate from actors.

Liminoid genres present a particularistic view, often an alternative view of the dominant social organization, and hence are capable of serving up social criticism (Turner 1982).

In the process of a successful performance, the individuality of the individual is muted and the actor sometimes achieves a state of flow. Such a flow is a transportation, but it can eventually lead to a transformation. Audiences of liminoid genres are temporarily affected during a performance. This is usually a mere transportation, though genres which allow for a good deal of audience participation probably also engender transformation.

It should be observed that this schema defines ritual as a performative genre in traditional agrarian societies, and theatre, as a performative genre in modern industrial societies.

The use of traditional theatre for development in Africa, however, clearly seems to envision a sort of hybrid between these two poles. Theatre for development advocates a kind of grassroots participatory activity wherein optionally chosen development themes are selected and enacted improvisionally by villagers (James 1988). To my knowledge, those who advocate it are only dimly aware of its liminal aspects and hence perhaps underestimate its transformative potential. The subsequent discussion hopes to shed some light on the subject.

Theatre for Development in Africa

In recent years, the body of literature promoting the use of village based drama as a tool for development has been substantial. Indeed, by the late 1970s popular theatre for development was being used and described in several countries including Nigeria, Zambia, Sierra Leone, Zimbabwe, Botswana, Malawi, Cameroon, and Kenya (Desai 1990).

Much of such literature which advocates the use of this genre of communication is enthusiastic about its democratic potential. It views village drama as an alternative to the top-down modes of development communications in which urban technocrats presume to know what is best for rural folks (Friere 1970, 1976; Kidd 1982; Malamah-Thomas 1985). Indeed, a number of scholars have focussed on the interactive capacity of village based drama, one which contains built-in feed-back mechanisms (Hall 1981). Similarly, others have argued that the use of theatre can promote active involvement, self-reliance, and participation in the development process while engendering critical consciousness (Byron and Kidd 1977).

Many scholars have lauded also the discourse style of village based drama noting that it is particularly effective because it speaks to the people in their own idiom. Wilson, Epskamp and Morrison, in their chapters, point out that drama, an oral form of communication, is highly tuned to the realities of African village life as do Pratt (1987), Ong (1982).

This analysis has no quarrel with the above formulations. It merely wishes to present a further argument: that the performance of village based theatre, which is considered here to be of the liminal variety, in traditional African societies can engender a ritual-like or liminal atmosphere. As such it can be seen as deeply serious, and the behaviors it advocates as somewhat less than optional. This village theatre can have transformative power; when transformation is the very essence of development. Here the definition of development I use is one borrowed from Gecau (1982: 89):

> The mobilization and transformation of men and women — a nation's most important resource — to become the makers, the subjects, and the objects of their own lives and history.

I should like to explore the Turner/Schechner ideas in conjunction with the presentation of one experiment in village based drama. I will attempt to show that in the enactment of one particular performance in rural Liberia, liminal elements were present and something near to liminal occurred.

Case Study

In 1989, the Liberian Rural Communications Network project (LRCN) launched a series of experiments designed to determine if village based drama could be organized within the listening radius of the stations and whether these could be adapted for radio. Before proceeding to the discussion of the drama itself, it is necessary to describe the LRCN.

Production Context of Drama
The Liberian Rural Communications Network was a development radio effort jointly sponsored by the Liberian government of the then President

Samuel Doe and the U.S. Agency for International Development. The project was conceived with a view to promoting development through indigenous-sounding broadcasts. The three LRCN radio stations, unlike the National Liberian Broadcasting Service (LBS) station, were located upcountry, in the cities of Gbarnga, Voinjama, and Zwedru.[7] These LRCN stations provided for the first time, radio broadcasts intended for Liberians beamed from outside the capital city of Monrovia.

The network programming was organized to promote a maximum of interaction with the listenership. Audiences were encouraged to form listener groups to meet and discuss the programmes, to raise funds, and to help support the stations, and to provide feedback through a variety of channels.

The content of the programming consisted largely of development messages derived locally or in Monrovia depending upon the funding agency involved.[8] Other content consisted of local news and entertainment broadcasts, including a good deal of traditional music and oral histories recorded within the listening areas.

The most innovative aspect of the LRCN project was the fact that it broadcast in local languages — twelve Liberian languages and English. Producers had been chosen for their ability to express themselves in both English and a language native to the listening area in question. When using the vernaculars, producers were encouraged to express themselves in the full idiom of the language. They were discouraged from allowing the linguistic biases of English to creep into their announcing. Furthermore, elders well schooled in local languages were called upon to monitor the broadcasts to insure linguistic purity. It was within this production context that the village based dramas were conceived.

Creating and Enacting the Drama

The project engaged the services of a Washington based radio drama consultant and those of Liberian dramatists trained in western theatre techniques.[9] They were expected to combine their efforts to teach the LRCN radio producers how to organize and mount village drama productions and record them for radio. To our knowledge, this had not been done in Liberia, certainly at least not for radio, and certainly not by our producers.[10]

The project began at the Zwedru radio station, in south-central Liberia, in the heart of Krahn country. Here the consultant and the dramatist met on the first day with producers to work out a story for village production. The producers selected a theme from one of the station's ongoing development campaigns, — the promotion of wet rice agriculture or "swamp rice" over upland rice.[11] Once a topic was selected, the group discussed the advantages of swamp rice and the disadvantages of upland, and then began to develop a storyline.

As they considered how they would dramatize the upland farming message, they began to create characters. The producers decided to present rain, a tree and a groundhog as talking characters who would warn the farmer that he should leave the forest (where upland rice farming is underway using traditional slash and burn methods). They also turned the swamp into a fertility character who beckoned with her rich bounty. Her name and the name of the play was thus "Mother Swamp." The basic story framework as conceived by the producers is provided as follows:[12]

"MOTHER SWAMP"

Drums with Robert's Intro: (Scripted studio portion indicating the origin of the drama and the names of the players.)

Opening Chorus: *"This message is for You!"*

Farmer cries for the rain to come. The rain teases the farmer. *The village echoes sound effects.* The rain allows a little crop to grow, but not much.

The villagers are tending the small crop, but when their backs are turned, the groundhog comes. The farmer chases him. The groundhog tells the farmer, "We are all here to eat." The village reacts to the groundhog and *echoes the sound effect.*

The chorus sings a hunger song. The small small harvest was not enough to keep them well. But then the new season comes and they begin to brush the field and fell the trees on the new land they are moving to. *The Chorus sings this and makes the sound of brushing.*

The farmer is cutting the tree. The tree and the farmer discuss. *The sound of cutting is echoed by the village. The village enlarges the sound of the sound of the falling tree.*

The wife arrives with lunch for her husband. She finds him moaning and hurt. He describes to her what he is thinking. The rain teased him. The groundhog robbed him. The tree warned him. What did this all mean? Then he dies.[13]

The wife runs to town to tell the village. She meets Mother Swamp on the way. *Mother Swamp and her chorus describe the bounty they offer.*

The wife goes to the chief. The chief orders the town crier to gather the village. *The town crier goes from microphone to microphone* calling the villagers who respond. They receive the news and leave for the body.

On the way they again meet Mother Swamp. She sings her song and *it is followed by drums.*

The chief is approached by the swamp farmer who explains the benefits of swamp farming. The chief and his elders leave for the Oracle.

The Chorus sings a "Going to the Oracle" song.

The chief and the Oracle discuss and the chief goes back to the village.

The Chorus sings a "Coming back from the Oracle" song in anticipation of the outcome.

The town crier gathers the village for the conclusion.

The chief describes all that has happened and the message from the Oracle. The rain's and the ground hog's and the tree's warnings are repeated and the summation.

The chorus sings a "Good Fortune" song. The chief concludes and the villagers leave for Mother Swamp with their rakes and hoes and tools.

Mother Swamp sings her welcome. The villagers answer with their promise to learn her secrets well.

Drums with Robert's Conclusion; (Scripted portion thanking participants, and urging the adoption of Swamp Rice Agriculture).

     After conceiving the story, the producers assumed characters and role played the parts in English. They taped the production in the studio to see how it flowed.

     That evening the production team travelled north to the selected and very rural Krahn village of Gwemble, located about ten miles off the tarmac road which ends another ten miles outside of Zwedru. When they arrived, a group of villagers was waiting for them.[14] So after a sharing of proverbial cola nuts,[15] the team set to work. They began by presenting the villagers with the storyline, and asking the folks for their suggestions and comments. The LRCN team then asked villagers to volunteer to assume roles. This was accomplished with a surprising ease. Participants assumed age appropriate roles and they provided a rendition of their parts. The group selected Mother Swamp and a women's chorus formed around her.[16] The chorus produced sound effects and sang appropriate warning songs. Quite satisfied with their efforts the team departed, asking the actors to work on their roles before the next evening.

The next night the producers returned for a taped rehearsal. They explained the story again. There was one scene in which the farmer character (who persists in his upland farming practices) was to die when a tree fell upon him. The villagers were reluctant to play this scene as it had been written and argued that it be changed.[17] The Liberian dramatist felt that the integrity of the play would be destroyed with such an alteration. So he urged them to play the scene as it had been conceived. They did so but with great hesitancy.

So with a minimum help from the production team, (who were preoccupied with the microphones and the mixing board) the villagers launched into the enactment of the play. The consultant reports that "the singers performed energetically and the actors brought enthusiasm to their roles" (Brooke, 1989; 4).

There was also widespread participation in the drama effort in Gwemble. Actors included elderly and important personages, women of childbearing years, young men in their twenties, and even children.

Somewhere before the end of the play, the tape ran out, but by then, the actors, the chorus members including many children as well as many of the radio producers and technicians had gotten into the flow and not one of the LRCN staffers had the heart to stop the performing villagers. So the group got through the entire performance (Brooke 1989). The death scene, however, was very flawed. It engendered giggles and nervousness from the performers and audience alike as it was performed.

Still, at the end of this second evening, the producers felt the exercise had gone well. They left the village late that night proud and pleased with their efforts. They promised to return with fresh tape (and more of it) so as to do a final taping including the missing final scenes.

Alas, the next night, the final taping was marred by a number of significant events. The producers arrived at Gwemble to find few of the previous night's performers. Attempts to summon a cast involved a good deal of waiting (more than is usual in a village structured chronemically according to natural rhythms) and finally some substitutions. Explanations regarding the missing cast, especially Mother Swamp were proffered. Some said the women's husbands had been annoyed because their previous night's participation had taken them from their chores. Others said the women feared their participation in this effort would elicit ridicule (or worse) from neighbouring villages. Why a change of heart on this point was made clear somewhat later, after the consultants had left Zwedru.

It was obvious, in any case, that the death scene had been a major bone of contention. When a second cast was finally pieced together, the group insisted the death become an injury, one from which the farmer would recover. On this third night, the night of final taping, it was evident, that flow was not achieved. The drama consultant reports on the issue:

> The performance was okay, but the enthusiasm and energy of the previous nights were gone and everyone was relieved when the show was over (Brooke 1989: 5).

And the musical performance that third night was also poor (Pratt 1989: 2).

Aftermath and Analysis
It was later learned that sometime before the production, there had been a palaver (an argument) between a Gwemble farmer and one from a nearby village. From what the LRCN producers could piece together, the altercation had been over methods of rice production. So when people from this second village heard of the Gwemble "dress rehearsal" drama rendition, they were deeply offended. Some even speculated that the villagers of Gwemble were performing a curse.

The story of the Gwemble village drama illustrates several important points. On a very fundamental level, it established that village based drama production was workable.[18] From the outset of the exercise, the producers were keen on the concept. Most thought it a worthwhile exercise. They demonstrated their initial enthusiasm both in the creative session at the station and in their willingness to travel to Gwemble three nights in succession and by working well beyond midnight on the latter two occasions.

In the matter of dramatic enactment, most seemed to know intuitively how to proceed. And because of their familiarity with the development projects and issues they could quickly select a theme. Furthermore, once the theme of the story was selected, they slid rather effortlessly into the construction of a storyline.

The villagers were also keen and cooperated to the maximum, sharing their resources of time and energy, their dramatic style and musical talents and their overall enthusiasm. On a deeper level, the exercise in the production of village-based drama churned up some fairly liminal elements. These merit further examination in the light of what has already been explained about liminal/liminoid genres.

Use of Liminal Symbols

In the construction of the elements of the drama, producers displayed a tendency to anthropomorphize or personify by giving life or a spirit to natural phenomena — the wind, the rain, the groundhog. The swamp herself was deified into a bounteous earth goddess. The creation of an earth goddess seems very significant in this village which practices swidden agriculture.[19]

A Social Drama

Another noteworthy aspect of the drama created was its approximation of a social drama, the conflict between life and death for the villagers. Indeed the metaphor of the production is clear. The cultivation of upland rice equals death; swamp rice cultivation will bring life. Also noteworthy was the structure of the play which contained four phases: breach — (the farmer fails to heed the warnings of nature); crisis (the farmer lies wounded pondering the advice he has ignored); redress (the villagers consult the oracle); and reintegration (the villagers take up swamp rice farming).

Group Consciousness

It should be noted that even though criticism and restructuring of the storyline was invited by the producers, little was offered by the villagers. (The matter of the death scene was an exception in this regard. Its discussion is taken up below.) This suggests that either the group was strongly cooriented around the elements in the story, or that they had been ordered to do so by their own chief. Either way, the lack of divergent opinion regarding the storyline suggests the presence of a strong element of group consciousness.

Also significant in the enactment was the relative ease with which villagers assumed and slipped into roles. This was done with little apparent disagreement or palaver. The consultant discusses the issue in her report:

Drama roles were assumed by age appropriate people (the Chief and the Town Crier played themselves), an old aunty was played by an older woman, and so on (Brooke 1989: 9).

Indeed this tendency has been noted in at least two other works on rural theatre in Africa by Boal (1979) and later Fabian (1990). Turner's formulations shed light on this particular aspect of the performance; especially his assertion that African villagers can easily assume roles in ritual activities because traditional African social life is bound up in the assumption of a myriad of social roles.

In the simpler, pre-industrial societies acting a role and exemplifying a status was so much a part of everyday life that the ritual playing of a role, even if it was a different role from that played in mundane life, was the same kind as the one played as son, daughter, headman, shaman, mother, chief, or Queen sister (Turner 1982: 115).

Turner (1982: 115) explains further that the indicative and the subjunctive moods are not very far apart in pre-industrial society:

The difference between ordinary and ritual (or extraordinary) life was mainly a matter of framing and quantity, not of quality. In ritual, roles were separated from their embodiment in the ongoing flow of social life and singled out for special attention, or else they were seen as points of entry and exit on a continuous process (boy to man; girl to woman, commoner to chief, ordinary village member to member of a hunting cult, ghost to ancestor, etc.). . . . But in these societies, acting was mainly role-playing. The persona [assumed character] was the dominant criterion of individual identity. Thus, the great collective which articulated personae in hierarchies or segmentary structures was the real protagonist, both in life and in ritual.[CITATION?!]

Indeed, the very speed in which actors and actresses roles were selected by the villagers suggests a great deal of organic solidarity and a great deal of agreement about who was to be whom.

Participation Required? Work Play-Continuum?

It should be noted that participation in the drama activity was said to be voluntary, hence liminiod. Indeed, it must have been as on the third night of taping, the cast was different from that of the other two. The composition of the cast on the second night was particularly intriguing (This is the night when the liminal elements occurred). It was noted that some of the characters played themselves. This was true of the town crier and the chief.

It was also suspected that the women's chorus which was assembled on the second night was the customary religious chorus. This suggests that participation may have been less than optional. Certainly no one was coerced to play, but it does seem likely that many of the personages in the village felt called upon to "do their customary thing." This sense of social obligation bound up in the immediacy of the moment may also help explain why the women were later said to have neglected their chores.

Entire Community Involved? Audience Participation?

The keenness of the villagers on the second night, the night when flow was achieved, was evident to all who participated in the exercise. The tone of the consultant's report suggests that the performance was somehow woven into the fabric of the life of the village. Elsewhere she described children wandering in and out of the drama circle participating as they saw fit. Her discussion of this community event is enlightening.

> The villagers cooperated with maximum enthusiasm and inventiveness. They coached each other in using microphones properly; they passed babies through the drama circle when the sound of children crying was needed; they enriched the scenes by adding emotion, dramatic tension, and dialogue. Women nursed their babies as they performed. The crowd responded appropriately when the crowd scenes occurred; the songs were performed in a lively way (Brooke 1989:10).

Also noteworthy in this context was the difference in the performative experience between Gwemble and that of other villages located closer to main roads. In these latter cases villagers regarded the village-based dramas proposed to them by the LRCN producers as an activity to be delegated to the cultural troop of the village in question. Teenagers and persons in their twenties predominated such groups. And none of the older folks came out (Brooke 1989). These differences highlight the lack of specialization and role segmentation in Gwemble and suggest that something approaching a group process typical of the liminal experience was approached in the Gwemble village drama experience.

Deadly Serious?

From the points noted above, it should be apparent that the more solid and organic the village, the more likely was the content of the drama to be taken seriously. It is one thing for neighbours to ignore the fantasies of teenagers engaging in a drama troop. It is quite another to take lightly a participation involving elders including such recognized personages as the chief and his messenger and the women's chorus.

In such a context, a reference to the death of a slash and burn farmer, particularly where altercations over farming had recently occurred between neighbours, seemed dreadful indeed. The likelihood that the death was probably seen as inauspicious must only have compounded the liminality of the performance. The hesitancy of the villages in Gwemble to perform the death scene clearly suggests the seriousness with which they regarded this matter. It certainly suggests a shared belief on the part of the villagers in sympathetic magic. Clearly the villagers themselves recognized the capacity of performative genres to bring about certain ends. Perhaps they had recognized the deep structures of the drama. And later on they asked that the tape not be broadcast because a serious conflict had broken out between them and a neighbouring village.

Transformation?

The drama at Gwemble seems to have revived an existing division between Gwemble and a nearby village. That a rift now grew between the folks of Gwemble and their neighbours suggested that the play was indeed transformative. Things had been different before the production team travelled to the village! Indeed, the enactment of the second appeared to have engendered its own social drama!

How the altercation was handled in the aftermath could help to tell us to what degree the activity was liminal or liminoid. If for example it had served to integrate the entire community of Gwemble (say if everyone in Gwemble bonded together against their angry neighbours), it would appear more liminal. If it served, on the other hand, as a mechanism for social criticism (for those who practice one type of farming or another; or on meta level, for those who participate with radio producers in village-based drama!) it would seem more liminoid.

At another level of analysis, it is appropriate to ask the extent to which this sort of transformation is of the sort hoped for or anticipated by the development planners. Indeed the answer would seem to be a resounding "No!" It is likely that for the average planner, such a conflict would seem irrelevant, annoying, and quite "besides the point." That meanings emanating from an urban centre, albeit a nearby one, should engender such local conflicts only shows us how much psychic distance there is to travel for the integration of rural African villages with one another and with the larger urban agglomerations.

This is not to say that the project was unsuccessful. Perhaps the consciousness of the LRCN team was at some variance from that of the villagers of Gwemble and that of their neighbours. Still, metaphorically, all spoke the same language. The performance made sense to the villagers and they were able to perform it well. The performative piece was multi-layered in meaning, in the best tradition of the genre.

There is no reason to doubt that other changes advocated by development planners will engender further conflict among small social groupings. And however inconsequential or overblown these conflicts will appear to the planners, they will be very real to the protagonists. These conflicts, or social dramas, will require working out and through if development is to proceed. What better means can be suggested for this process than a theatrical genre which itself is derived from social drama?

The ideal village based drama would then be one conceived of in the village itself. Two villages which found themselves in conflict over farming issues might resolve their differences through drama while "subjunctively" trying out new farming methods before the ultimate agricultural transformation.

What is envisioned here for village-based drama is a country filled with rings of villages all working out differences through performance. The rings should gradually enlarge as the villages became better integrated, moving toward larger and larger units. Eventually these rings should encompass the capital city!

This is a model for "bottom-up" and shared communication. It is utopian indeed! In this vision of balanced and interactive development, the logico-meaningful modes of integration are still in tension with the causal functional modes. But the divergence is tolerable, challenging, even fun, rather than potentially threatening.

Alas, we live in a far less than ideal world, a world in which the funerals of Paijan threaten constantly to break down among people adjusting to urban loyalties, national political slogans, and contractual obligations. These people are still clinging to the rituals which gave meaning to the rural lives of their parents. But we also live in a world where development projects come from the EEC and where radio producers live in cities, even if these are small and rather rural by world standards. Because of the geographic, psychic, and social differences implied in these facts, hostilities may arise as villagers interpret enactments according to the rules of their own *weltanschauung*. An ignorance of these possibilities is both unwise and dangerous. It is tantamount to a failure to recognize the role of culture and its independent operation in the process of social change.

Final Thoughts

It should be clear at this juncture that ritual and its cousin theatre have tremendous transformative power. This chapter has sought to analyze this power by attempting to explain how it works. The degree to which villagers are allowed to form their own dramas and to choose their own themes will be the degree to which this genre is revolutionary. Obviously, if themes are chosen in capital cities by elites, the result of the theatre practice may well ultimately be the cementing of the status quo. Alternatively, if the least powerful in the villages are given the opportunity to select themes, a radical transformation of the villages and perhaps the nation might result.

Given the inherent plasticity of cultural systems and their performative genres, we cannot know what will be the outcome. And it does not behove us to make overdeterministic predictions based solely on the sources of thematic materials. What we do know is that in the enactment of village based drama in Africa, there lies a tremendous potential for change. "The philosophers have only interpreted the world in various ways. The point is to change it" (Marx 1934). I, like Marx, have provided an interpretation for the role of theatre in African development. Others are now called upon to make that change occur.

NOTES

1. See especially Gregory Bateson's *Steps to an Ecology of the Mind*; Mary Douglas' *Implicit Meanings* and her *Purity and Danger*; Clifford Geertz's the *Interpretation of Culture*, and Victor Turner's *The Ritual Process*. For later work, see James Peacock's *Rites of Modernization*.

2. See especially Daniel Lerner's *The Passing of Traditional Society* and Wilbur Schramm's *Mass Media and National Development*.

3. The evolution of the organizational structure of the LRCN is complex story involving first a set of negotiations between the U.S. AID and the government of President William Tolbert and subsequently a different set of agreements with the Doe government which seized power in 1980. For additional details, see Final Report of the Rural Information System Project.

4. For a useful discussion on this topic, see Beryl Bellman, *The Language of Secrecy*.

5. This definition is derived from Victor Turner who was greatly influenced by Auguste Compte, Montaigne, William Dilthey, Aristotle, Arnold van Gennep, Bronislav Malinowski, and Claude Levi-Strauss.

6. Adaptation for radio involves other technical considerations, both in terms of scripting structure and in terms of recording, These questions merit attention in a separate paper treating the issues involved in broadcasting a dramatic performance.

7. Gbarnga was located in central Liberia, in Kpelle country, in Bong County. Its population was about 14,000. Voinjama was located near the Guinee border in Lorma country. Its population was about 12,600. Zwedru was located near the Ivory Coast border in Krahn country. Its population was about 12,000. Figures have been taken from 1974 population figures and doubled, given urban population growth and general population growth in West Africa (Nelson 1985: 291)

8. The project accepted funds from sponsors, other development agencies who would buy time on the station(s) to promote their projects. For national campaigns, they would often contract for time on all three stations. For regional or local campaigns the airwaves of only one of the three stations would be used.

9. The Washington based consultant had significant experience teaching performing and media arts in the Washington School system. In this connection, she had also produced radio dramas for National Public Radio. Her role was to teach the producers how to work story ideas into scripts, how to sequence dramas for radio, and how to compensate for the absence of the visual while using the aural channel to its maximum potential. The Liberian dramatist had considerable experience working with professional and improvisational groups in urban based settings both in the United States and in Liberia. He had also worked extensively as a producer-director in secondary school theatre in Liberia. His role was to organize and animate the actors for the production. Both consultants were working with non-literate African villagers as actors for the first time. It was hoped that through their combined efforts and talents, along with the aid of the LRCN production team, village based drama for the radio could be realized in Liberia.

10. A variety of urban based theatrical initiatives have been organized in Liberia over the years.

11. A project sponsored by the European Economic Community.

12. The storyline has been taken from the consultant's report with only mechanical adaptation for presentation in this paper.

13. On the third evening, this scene was changed into an injury scene.

14. Appropriate arrangements had been made earlier that day.

15. A welcoming ritual showing social solidarity and amiability common in West Africa.

16. The women's chorus seems to have been the customary one normally called upon for performances in Gwemble.

17. In Gola-Kpelle cosmology, such a death is considered a "bad luck" death. See

Bourgault, 1990, p.12. A person who dies in this way is believed to be a victim of witchcraft. He is denied ritual burial. The death is apparently inauspicious in Krahn culture as well.

18. Producing dramas for radio, while engaging and intriguing is cumbersome. In non-electrified villages, a generator is needed to power the equipment. Typically, a generator is noisy and must be situated at considerable distance from the microphone. A number of other technical problems render the activity somewhat involved and complex.

19. The Krahn of Liberia have been little studied, and little is known about their cosmology. Linguistically, they are part of the Krou group which includes the Beté, Dida, Guéré, Wobé and other groups of the Ivory Coast (Handloff 1991: 57). It seems likely that an earth spirit must figure prominently in these groups. Ethnographic accounts of such groups may be helpful in this regard. See Dekeyser, P. and Holas, B., *Mission dans l'est Libérien* and Girard, J. *Dynamique de la Societé ouebé.*

Chapter 7

FORUM THEATRE: A CULTURAL FORM OF COMMUNICATION

Joy Morrison

Communication, from the Latin *communicare*, means to share. In its true sense then, communication ought to be a two-way exchange of a human nature. It is a give-and-take whereby both, or all, parties are enriched. Mass communication is something quite different. It usually involves a medium of one form or another. There is no sharing, merely a mass transfer of information. There is a source of the information, and there are unidentified receivers out there somewhere, among the masses. African peoples, particularly in West Africa, share their communications orally, face to face and have shown a preference for this form of communication (Morrison 1989a; Pratt 1987).

Oral and personal communication plays an important structural role in African societies, it is the glue that bonds people and communities. The appropriateness of these traditional forms of communication — as well as the inappropriateness of Western mass media — in Africa has been well-documented recently by African scholars[1] and as far back as 1961 by an American scholar.[2] Communication with rural people via the mass media is often problematic in Africa for a variety of reasons. It could be that the language used, for example, English or French, is not spoken by many in the population or that many people are illiterate or that many do not have the receivers necessary to capture the signal. In my research, I was told that these media only entertain and distract, and are not taken seriously (Morrison 1989a). Modernizing elements have attempted through many different means to supersede the personal communication with forms of mass "communication," or mass transfers of information, for the sake of efficiency. This is a one-way activity, depriving the receiver of any opportunity to participate. Thankfully, in most parts of rural Africa, these attempts have been resisted and thwarted by the very factors they have sought to overcome or change — those sometimes conservative forces of tradition.

In this chapter, I will present a particular form of communication in West Africa which appears to fit well with traditions, and explain how it functions. In the decade of the eighties, the government of Burkina Faso attempted to communicate family planning information to the population: a typical case of social change and its handmaiden, development communication. The population growth rate of 5 percent was a cause for concern for the government. The high birth rates were also resulting in declining health of women, and a high infant mortality rate. For a number of years, this information was transferred through radio spots broadcast throughout the country, pamphlets distributed at clinics, and billboards in the cities and larger towns. The campaign did not have much success and population figures soared in the rural areas. Midway through the decade, the government tried another vehicle — an indigenous theatre group, the Atelier Theatre Burkinabe (ATB). The form of theatre practised by this group is particularly appropriate in a primarily oral culture — that of forum theatre. It is a participatory form of performance modeled on some of the most ancient dramatic genres.[3] The ATB began working with rural people in 1979 creating sketches to assist in rural development programmes. The forum theatre format was selected for several reasons: its oral nature, the significance of performance in the culture, the role of cultural forms in social learning, and the opportunities afforded in the forum theatre format for dialogue.

Many forms of communication may be employed to convey meaning and messages. In Africa, as in other parts of the world, drama is a popular cultural genre and is often utilized as a communication medium. Both Bourgault and Epskamp articulate examples of indigenous drama as vehicles for communication. Its use of language, symbols and styles familiar to the audience may contribute greatly to effective communication. The communication context is crucial, and it is my contention that participatory performance is at the heart of this interactive communication process, as performance is an important structural component of most African cultures. Colletta (1977: 14) feels that existing folk culture "placed in a structural-functional framework, can be identified, mobilized and used to carry development messages through the sensitive modification of their multiple functions." Pratt (1987: 13) adds that "to the extent that any media are effective in reaching this target population at all, it is likely to be such traditional channels as folklore, drama, and puppetry which most effectively deliver development information in rural areas."

The viewpoints expressed by African scholars such as Odeani, Obeng-Quaidoo, Pratt, Mlama, Ugboajah and Boafo support my position that African communication should be studied in its specific cultural context. Their perspective is one of taking the cultural framework of society into consideration when conducting research.

The Oral Nature of Traditional Communication

Oral communication remains vital in Africa today. The oral tradition persists because Africans are still largely illiterate and, according to Obeichina (1975), most live in traditional, and culturally and linguistically homogeneous village settings which foster oral culture. Laye (1984: 26) puts it well in his book *Guardian of the Word* when he says:

> We touch upon one of the fundamental aspects of the African soul: the word, the love of palaver and dialogue, the rhythm of talk, that love of speech that can keep the old men a whole month under the palaver tree settling some dispute — that is what really characterizes the African peoples.

This love of words and speech manifests itself in many ways in Africa. In Burkina Faso, as in other African countries, there is an aesthetic and a respect for speech that is inscribed in the culture. The art of rhetoric constitutes one of the artistic forms most appreciated in the tradition. Burkinabe *griots* (praise-singers and story-tellers) have for centuries transmitted news and information and have influence and importance. *Griots* are valued and create systems of discourse which instruct and entertain. They are also the record keepers, chronicling births, deaths, and the history of relationships within and outside the community. In his or her stories and tales the *griot* uses many proverbs — powerful words which form part of the common collective consciousness. Siqwana-Ndulo (1989: 22) writes that "the proverb validates and augments a trend of argumentation, affirming to the discourse participants that the speaker's viewpoint has the blessings of an unquestionable truism."

Ong (1982) considers that people in oral cultures learn by apprenticeship, by listening, by repeating what they hear, by mastering proverbs and ways of combining and recombining them, by assimilating other formulary materials, and by participating in a kind of corporate retrospection. In an oral culture, conceptualized knowledge needs to be repeated often or it can vanish. Ong (1982: 41) feels that "this need establishes a highly traditionalist or conservative mind-set that with good reason inhibits intellectual experimentation." However, when attempting to bring new ideas to rural cultures, the stored knowledge is being added to or adjusted; this needs to be done incrementally. I have suggested that this way of learning — orally and incrementally — is not understood by Western communication scholars, whose own theories of learning and media effects may be different (Morrison 1991).

The Role of Performance in Culture and Social Learning

Performance is central to all African cultures. It includes visual aspects such as costumes, dances and symbols; oral aspects such as songs, narrative, poetry and proverbs; and spiritual aspects such as deities, tricksters, animism, and other belief systems. Performance serves to order and control society, and can communicate much through expressive actions. According to Stone (1986), all African adults learn to perform with some degree of proficiency in song and dance. To be part of a community and to be a social being is to know about performance, be it drama, or music or dance. Performance and traditional media are grounded in the metaphors of an indigenous culture produced and consumed by members of a group. They reinforce the group's values by teaching and initiating members of a society.

In most Third World cultures, oral traditions and cultural performances are the primary ways of educating the young and of promoting and reaffirming beliefs and values among adults. An oral medium such as dramatic performance is expected to "teach." Storytelling, drama and songs are some of the ways of learning and instruction in an oral tradition. Access to complicated or abstract ideas through narrative form, in which characters act out common human problems, is not unlike the popular televised soap operas in industrialized countries. A medium designed to teach is more acceptable when it uses national languages, avoids vertical elitist techniques of communication, and encourages egalitarian methods such as two-way communication with feedback. Stone (1986) states that Africans often think of performance in a transactional sense, like two people pulling at either end of a tug-of-war rope. Rather than two people standing alone, one person rarely exists without the other. Forum theatre reflects these African realities and avoids the eurocentric form of theatre which is strongly hierarchical. Audiences are incorporated through an "in-the-round" performance format.

UNESCO researchers have studied the use of the theatre as an educational tool and as a medium promoting information flow, and they found certain features which assist communication. A UNESCO official observes in a document on the use of drama for social change that

> the performing arts provide a particularly good meeting point for traditional values and the requirements of progress. Their form is by no means set in rigid structures; the strict requirements of oral tradition and of handing down from memory necessitate social supervision of creation and rule out too personal individual performance. Their traditional content is subject to constant change, as the experience of each generation is slowly passed on to the next, fulfilling a social purpose (Anon 1975: 18).

Forum Theatre

The theories of education and political activism of Brazilians Paulo Freire and Augusto Boal have strongly influenced the development of forum theatre. Boal originated a genre of theatre in the 1960s in Brazil with his street performances for political consciousness-raising, and it became known as forum theatre. Boal's (1979) notion of "theatre of the oppressed" is essentially militant theatre with the objective of getting people to become aware of their own situation, to take charge of their own destiny, and to create an internal dynamic. Leis (1979: 13) describes it as "theatre in ferment, in which the participation of the public creates the work performed." Freire (1974: 13) felt that every human being, no matter how "ignorant" or submerged in the "culture of silence," is capable of looking critically at his or her world in a dialogical encounter with others and that "provided with the proper tools for such an encounter, he can gradually perceive his personal and social reality as well as the contradictions in it, become conscious of his own perception of that reality, and deal critically with it." A medium that permits this kind of encounter and communication could be very empowering for rural people.

Kompaore (1989), who brought forum theatre to Burkina Faso, claims that the fundamental principle of forum theatre is the non-dichotomous relationship between actor and spectator; the spectator can participate in the performance. Thus the feelings, opinions, and reactions of audiences are solicited and received in a non-threatening environment. Forum theatre has been used to communicate social development information such as nutrition, health care, and reforestation issues to Burkinabe people since 1981. In 1989, the Atelier Theatre Burkinabe took forum theatre to 21 villages to begin a dialogue about family planning. Their play was called *Fatouma: The Baby Machine.*

There are three parts to a forum theatre performance. In the first phase, the actors perform a play as an "anti-model," during which social problems are depicted in ways that may displease the audience. The problems are portrayed in a way that results in spectators desiring to change things.

In the second phase, the play is reenacted, scene by scene. Spectators are invited to take the stage and intervene as actors to propose changes and improvements to scenes they did not like, to engage in role playing, and to provide commentary. If the other spectators agree with the recreation, the person is cheered, and we proceed to another scene and another role. This solution implements the choice of the people. According to Kompaore, inviting spectators to come on stage and present positive social roles allows them "to project to an ideal future, a modified present and constitutes a rehearsal of social transformation" (Kompaore 1989). This is also important in that it permits local people to take the stage and vocalize their resentment in the guise of acting. This input from the audience helps project coordinators take into account the needs of the people in their planning.

The third phase creates the opportunity for a verbal exchange between the audience, actors, and health officials. The aim is to clarify the information. Audience members put questions to the actors and to specialists on the topic of family planning, and receive authoritative responses. This permits health care officials to give more details and also to hear about other public preoccupations. Forum theatre is thus not just entertainment — there's laughter, there are emotions — but they keep the artistic character of the performance. Participation is not solicited just to have speeches on stage by audience members; it is to raise awareness through art or artistic performance and to transfer information — in this case, contraceptive options — that empowers people.

Kompaore (1989) feels that allowing the public to participate in theatre is merely conforming to the desire of the people. This is proven by the fact that "whenever we give the floor to the public, they express their opinions spontaneously. We never have to force, to coerce anyone to participate in our forum" (Kompaore 1989). Audience input is solicited and received, and this two-way flow contributes greatly to its acceptability as a medium of communication.

The role of the forum moderator is a crucial one. He or she acts as a facilitator to ensure that all viewpoints are given equal opportunity of being heard. This role has been compared to that of a moderator in a focus group discussion, where a tactful assertiveness is needed. Participants must feel comfortable airing views that may seem unpopular, and accommodated without ridicule. An important element here is the participation of all. When an important national issue such as family planning is debated people have a sense of commitment at the national level, and a feeling of being directly associated with policy formulation at a policy level (Semana 1984).

Another strength of this kind of performance is the use of the indigenous languages wherever possible. Moore is the language spoken by the Mossi people in Burkina Faso. Actors in the group are selected for their Moore language abilities and knowledge because the village audiences appreciate a good command of the language, its nuances and its proverbs.[4] An ability to use these proverbs judiciously is highly valued. They are so powerful in the culture of the Mossi people that when employed in an argument, a person wins power over an opponent, because, as these sayings originate with the ancestors, they cannot be refuted. Many in the audiences found expression for some of their resistance to family planning in their proverbs. An example is the Moore version of the maxim "God gives, and only God can take away." Health care officials have to have quick responses, or their programme loses its credibility because of the power of the proverbs. A proverb I heard in response was "if you have a chicken with too many chicks, many will die but if she has only a few, they will be healthy and grow."

The play *Fatouma: la machine aux enfants* (Fatouma, the baby machine) was a collective creation of the Atelier Theatre Burkinabe and concerns Fatouma and Tanga, a young couple who marries and produce a child who is welcomed and celebrated. A second child is born the next year and a third the following year. After eight years and eight children, the marriage is in ruins due to problems created by Fatouma's refusing Tanga's sexual advances. In addition, the children are in rags and there is never enough food. However, Tanga's father convinces him to continue enlarging his family because of the extra labour provided by children and the status gained by large families. In addition, the grandfather wishes to have many descendants in order that he may be remembered after he is gone. The spectre of AIDS arises when Tanga begins to seek out prostitutes to satisfy his sexual needs. Then a health care worker visits Fatouma and shows her various contraceptive options. But her husband refuses to discuss these with her. Instead he seeks a contraceptive belt from a local healer for her to wear during their sexual relations. When she becomes pregnant yet again, he accuses her of infidelity as he is convinced of the effectiveness of the belt which she wore each time they had sex. He expels her from his house, publicly labelling her as a whore. The play ends with her expulsion from the family compound, a physically weak and broken woman.

Each performance of this play elicited very emotional responses from the audiences which were acted out in the forum. The most common and overt topic of controversy was the "generation gap." Young men came forward willingly to replay the role of Tanga and to try and deal with the pressure from his father. Elderly men recreating the role of the father state their reasons for wanting many grandchildren, and then the debate began. Although not as overt, gender tensions were played out, and there was never any lack of input from the audience. Young women, usually reluctant at first, recreated the role of Fatouma, trying to convince Tanga to allow a period of rest between births and to limit family size with the aid of contraception. The role of Fatouma created a great deal of emotion, particularly among the women. I noticed a number of women wiping their eyes and blowing their noses at the end of each performance of the play.

Older men in the audience were eager to replay the part of Tanga's father to explain his position as the patriarch and to emphasize the need for large families. At each forum at least one elder did this, and in the recreated scenes they were treated respectfully by younger men, who nevertheless disagreed with their views. Younger women seldom argued with an older man recreating the role of Tanga's father, but in one case a young woman replayed the role of Fatouma introducing a condom to Tanga, whose role was being replayed by an older man. She appeared to have difficulty convincing him, as he brought his convictions as an elder, not as a young man, to the scene. Even though he was playing the role of a young man, his stated convictions resulted in large numbers of young men, one at a time, coming on-stage to debate him on the topic.

The actor who played Tanga usually replayed his role, steadfastly holding his ground in refusing the pleas and reasoning from the women playing his wife. More and more women from the audiences became involved, creating interesting discussion at each performance. Certain scenes were designed to provoke the women, especially one where Fatouma is trying to feed her three children and a baby when Tanga arrives home to eat and have sex. In the play Fatouma gives him the food for the children and then asks the children to watch the baby while she obediently goes into the bedroom with her husband. In the forum that followed each performance, women in the audiences wanted to recreate this particular scene. Some women flatly refused his sexual advances, others tried to discuss the possibility of another pregnancy which could result, and still others complained that there was already lack of food for the children. One woman actually refused to listen to him until he provided enough chicken for her and the children, stating that she "did not have the energy for sex, but a chicken could help her energy level." The audiences were all very amused by these situations, especially the women, who seemed encouraged by each other at every forum. Woman after woman recreated this scene, giving it a different spin each time. One woman aggressively brought up the topic of contraceptives, provoking an "angry" response from Tanga. Another woman also asked that they consider some form of family planning as "it would improve our sexual relations as I would not be worrying about another child." Audience reactions to these scenes were indicative of the general feeling. Women approved wholeheartedly of these discussions judging by their laughter, applause, and encouragement of the female participants. The forums were usually the same length as the play (50 minutes) or longer, and were limited only by the actors' endurance.

The women contributed a great deal to the forum, slowly at first but with more and more courage as the forum continued. The performance appeared to give them a rare public platform to air their feelings, and there seemed to be a solidarity between women in the audience. Audience members taking various positions vis-a-vis the problem exposure express their divergent feelings and opinions, and this feedback provides a synthesis of ideas which emerges as a system of dominant thought. Forum theatre appears to meet the needs of a feedback mechanism and enables older people with the authority of experience to voice their opinions and have an audience of younger people as well as government officials. Because older people are the traditional teachers, this format is considered appropriate as a means of transferring information to village people. However, the older generation are also informed about new ways of solving problems in a non-threatening environment. Women especially seem to appreciate having a forum to air problems that are gender-specific, such as the topic of this performance. Forum theatre has an agenda-setting role as these issues are discussed and debated in the villages where the plays are performed long after the group departs. Family planning is a highly charged topic in Africa just as much as in the United States, both socially and politically. The play's content violated traditional norms by discussing sex, which is one of its aims. The forum resulted in a transition of a private topic such as sexuality to the public realm. This agenda-formation could have resulted in families discussing this sensitive topic, which had not been tolerated prior to the forum.

As an extension to its agenda-setting capacity, forum theatre, as a form of theatre for development, has been accused of coopting a cultural artifact — that of performance — and using it to manipulate rural people. The accusation has come from Western scholars (Colletta 1977; Kivikuru 1989; Kidd 1983). However, I saw it as a forum for discussion, a site for dialogue between the people and their government, or between traditionalists and modernists. I saw each party's input being heard and respected. I saw the art of rhetoric at its best, and artistic expression, and true communication. In interviews with villagers I heard nothing but praise for the government's efforts to solicit input from them, and at the same time provide an entertaining and educating experience.

Summary

Forum theatre is a mechanism for using a traditional way of communicating to solve a modern problem — that of family planning. When the views of the people are solicited and heeded, villagers know that they are active players in social change projects, and not merely recipients. In interviews with audiences young people indicated that they learned something by watching this performance, and older people were informed of something new and of its benefits. Younger people were keen that older people be convinced of the utility of contraceptives so that their use would be condoned. Although some young people seemed intolerant of the elders' views when they perceived these views as hindering progress, they did nevertheless recognize the hierarchy of authority. This notion of elder authority, a kind of gerontocratic practice, was raised in many of my interviews, as elders are responsible for much of the informal learning in villages. Younger audience members felt that it was good to have the older people's views questioned in a public forum such as this performance provided, as there is a generation gap in opinions concerning family planning.

My interviews led me to conclude that Burkinabe people of all ages accept and appreciate a theatrical performance as a medium of education and communication. Their orality, their preferences for interpersonal communication, and the tradition of performance in the culture are the main reasons for this acceptance.

The major strength of forum theatre is its basis in the community of sharing ideas and opinions. The oral nature of communication in Africa, love of dialogue and proverbs, the uses of performance, and the community spirit that characterize African peoples have convinced me that it is necessary to focus on more indigenous communication forms. The desire to be pleasured by communication and the ability to identify with circumstances portrayed, as well as the expectation of interactivity and participation emerge as theoretical propositions upon which this conviction is based.

African countries need to explore their rich traditional communication means and incorporate more of these into social projects. Perhaps such a model of communication, consensus building and community could be employed beneficially in problem-solving in other countries.

NOTES

1. Ugboajah (1985, 1986), Boafo (1989), Odeani (1988), and Obeng-Quaidoo (1986) among others.

2. L. Doob (1961).

3. For example the Koteba is an ancient performance accompanied by music held all

over Burkina Faso one night after the harvest begins. Actors dress up in costumes and provide short sketches for the amusement of villagers (Kompaore, 1977).

4.   An elder commented to me that to misuse a proverb is scandalous.

Chapter 8

INDIGENOUS RESOURCES IN A GHANAIAN TOWN: POTENTIAL FOR HEALTH EDUCATION*

Marie Riley

Introduction

For the past twenty years, communication researchers and development planners have been re-examining the role of mass media in development, and appraising its successes and failures. Many of the initial assumptions — that the media had enormous potential to inform poor, rural, illiterate people about ways to improve their lives and to act on the information — have been challenged (Rogers 1976a; Diaz 1977; Hedebro 1982). Even Wilbur Schramm, one of the most influential communication scholars, admitted:

> I should have been more sceptical about the applicability of the Western model of development. I should have paid more attention to the problem of integrating mass media with local activity. Above all, I should have given more thought than I did to the social requirements and uncertainties of development and in particular to the cultural differences that make development almost necessarily different, culture, to culture, country to country (Dissanayake 1981: 220).

Attempts to address these concerns have led to an approach to development communication characterized by an emphasis on self-reliance, on "grassroots" participation, on the use of "appropriate" technology, and on the productive use of local resources. Indigenous communication systems as defined by Wang and Dissanayake (1984: 22) seem to fit this new approach. They are

> embedded in the culture . . . and still exist as a vital mode of communication in many parts of the world, presenting a certain degree of continuity, despite changes,

by Wang and Dissanayake (1984: 22) seems to fit this new approach. Such systems include drama, dance, songs and story-telling, as well as their context and settings.

---

* This chapter is republished from the original published under the same title in the *Howard Journal of Communications* (Spring 1993).

Several of the advantages of such systems were identified by Dissanayake (1977):

> They have a credibility, particularly among rural populations, that modern media often lack. Press, radio, television and cinema are often "alien and elitist in outlook," and those at the periphery of modern life tend to identify them with the centres of power.

They employ the idiom of the people and are readily intelligible to them. Even the poorest have access to indigenous media, whereas radio and television may be beyond their means and newspapers beyond their comprehension. Indigenous media demand active participation in the communication process — the audience is involved rather than passive.

This chapter is based on field work carried out in 1989–90 which attempted to determine the extent and variety of indigenous communication resources in one town in Ghana, West Africa. An inventory is presented of those resources considered to hold the most potential for enhancing community development, particularly in the area of health education. Suggestions as to how those resources might be mobilized and some factors working against such mobilization are also examined.

Health Care in Winneba

Winneba lies on the coast of the Gulf of Guinea, some forty miles west of the capital, Accra. The town of about 40,000 people and the district around it are similar in many ways to the area in Papua New Guinea where Abrams (1984) made his inventory of traditional communication systems. Both have the following characteristics:

> a coastal region, within easy reach of the national capital.
> a traditional economy based on fishing.
> a high proportion of young people enrolled in school.
> an area where the Christian church has had a strong influence.
> an area where, despite health facilities, the incidence of malnutrition and child mortality remains high.

Preventive health care in Winneba district was provided by a health management team made up of four divisions — maternal and child health and family planning, environmental health, nutrition education, and communicable disease control, including immunization. An immunization campaign against six childhood "killer diseases" — polio, measles, tuberculosis, diphtheria, whooping cough and tetanus — was part of a broader effort sponsored by UNICEF and WHO to promote sustainable child survival and development in developing countries. Health personnel were assisted in the exercise by committees made up of members of various "revolutionary organs," including a national women's movement, members of the local elected assemblies, the national teachers' association and the road transportation union. Launching Child Survival and Development Week in December, 1989, the Secretary for Mobilization and Social Welfare pointed out that health authorities do not have the resources nor the outreach to carry health messages to vulnerable groups at the grassroots, especially women and young expectant mothers in rural areas. These committees were, therefore, expected to formulate plans and strategies that would help create awareness and understanding among this particular audience.

Figures from the 1988 Ghana Demographic and Health Survey show that out of every 1,000 live births in the country, 77 babies die before the age of one year and 155 children die before five years. The high mortality rate is due mainly to vaccine-preventable diseases, malaria, respiratory tract infections, intestinal worms, diarrhoea, and poor environmental health (Vieta 1989). In September 1989, UNICEF commended Ghana for having achieved 61 per cent coverage in the expanded programme of immunization and pledged continuous support to enable the country to reach 80 percent coverage by 1990. Although there was considerable activity by health workers regarding immunization, and heavy media attention, the achievements that were announced might in some cases have been suspect. In the Eastern Region, it was discovered that although the 80 percent target had been reached, only 42 percent of the children had received the full course of shots. Yet vaccines like DPT, against diphtheria, pertussis and tetanus, have to be given three times at regular intervals for them to be effective (Mante 1989).

This lack of follow-up was a serious problem in several areas of health care. The officer in charge of communicable diseases in Winneba related it to the treatment for yaws and bilharzia. Both treatments must be given over a precise period of time, yet parents tended to bring their children to the clinic or health post on market day, when it was most convenient, and didn't return for the second shot. Two members of the health team indicated that often women were simply not convinced of how important it was to complete the immunization process. They presented themselves for the team's first visit, when rallied by the chief or assembly member, but because the team was in and out quickly there was not enough time for adequate explanation and persuasion. A slight fever or swelling, a normal occurrence after a vaccination, was enough to deter a mother from bringing her child again. An examination of the immunization team's itinerary in Winneba district from October to December, 1989, showed eight to ten villages *per day* were scheduled for visitation. Given that many of these villages were off the main road, the teams were facing a considerable challenge in providing the service delivery alone. The opportunity to adequately prepare the population on the significance of the immunization programme, and the consequences of not completing it, simply did not exist.[1]

Insufficient inputs and need for reinforcement were also identified as problems by the officer in charge of nutrition education. In most villages, members of the health team spoke to small groups. Posters that were displayed were taken down when the team left, as these were in short supply and were to be used again at the next stop. The timing of the visit could also be a problem. Most people were at their farms during the day. Farming, a principal source of income, was a priority. The evening was seen to be a more suitable time for education and discussion, but no incentive was offered to the team members in the form of food or travel allowance to encourage them to spend the night.

The health team needed what James Grant, executive director of UNICEF, called "partners in the alliance" for improved health care. A core of basic information — on immunization, on oral rehydration therapy, on family planning — exists and has scientific legitimacy.

Putting that body of information at the disposal of all families is a task as enormous as the rewards it offers. It is the great health challenge of our times. And to meet that challenge, it will be necessary to forge a new public health alliance, to stimulate a new and *permanent mobilization* of a wide range of conventional and unconventional resources in the cause of health. (Grant 1988:33)

Such a call for input from the community needs examination. How could local groups, skilled in indigenous communication forms, make a contribution? This question is addressed below.

Ceremony and Ritual in Winneba District

The field work undertaken in this study attempted to determine what forms of indigenous communication are still in current use in Winneba and district, and what potential they hold as adaptable media for development messages, particularly health care messages. If culture is the context of communication, then close attention to these cultural forms may yield clues to guide message designers. Ugboajah (1985) and Sonaike (1989) have both stressed the need to catalogue the essential features of such traditional communication systems, and to detail their origins and purposes, mode and format.

The forms presented here include a group of male singers and drummers, the *asafo*, indigenous to the coastal Fante and Effutu people, and a women's group called *adziwa* or *adenkum* who make music with rattles and gourds, and who are accomplished dancers. An indigenous ritual, the performance of puberty rites, is also included. Finally, proverbs, which the Nigerian novelist Chinua Achebe described as "the palm oil with which words are eaten" and which enhance speech and give it flavour, are examined as a communication form.

Although these forms are changing and their function is often largely ceremonial, the role of ceremony is significant in the culture. Ritual and a sense of occasion are strong cultural components. The importance of ritual in new undertakings was observed in the author's first meeting with members of the women's dance and singing group, *adziwa*. They poured libation before the performance, informing the ancestors of the house of what was going to take place. A foreigner had come to learn from them, and the ancestors were requested to support the occasion and help to bring it to a successful conclusion.

The importance of following a prescribed order was also evident. During an annual festival, when a new health post was being commissioned, the chief was flanked by his linguist and elders while he spoke. So too was the government secretary for health flanked by his deputies. Likewise, when the Canadian high commissioner stepped forward to speak (Canada had contributed to the construction of the health post) the only two Canadians present were ceremoniously ushered to her left and right.

The ceremony surrounding female puberty rites can be extensive, including a ritual bath, new clothes, jewellery and hair style for the initiate. Special foods are prepared, special prayers are said, and a procession makes its way throughout the town. The rites differed in character from the *asafo* which, as indicated by Ansu-Kyeremeh in Chapter 10, is an institution by itself but also has its own development potential and communication attributes.

Asafo: Traditional Warriors

The male groups of singers and drummers found among the Fante and Effutu people along the coast were traditionally companies of young warriors, called *asafo*. They were charged with protecting the town from external enemies, and with ensuring the interest of the people against a chief who might exceed the limits of his power. Although the defence of the society has been taken over by the state, *asafo* still carry considerable political and social influence in Winneba, and have the power to dismiss a chief. They also undertake development projects through communal labour in the community and can be mobilized by their captains for special projects such as cleaning the town's water tank.

It is in their role as singers and drummers that *asafo* members appear prominently in the community. They are the force behind the biggest and most important annual festival, *Aboakyer*, and they perform at funerals, weekly occasions in Winneba and ones of significant ritual. Their song texts refer to wars and brave deeds from the past, and praise songs celebrate the bravery of war captains and the good deeds of individuals. At the same time, those who have shown bad judgement may be insulted.

The *asafo* companies are colourful groups, each possessing distinguishing uniforms, headgear, flags and emblems. Distinctions are made between different grades of musicians on the basis of skill, knowledge of repertoire and leadership ability. The two most important musicians are the cantor and the drummer. The cantor must have a clear, pleasant voice, must be able to remember the verses of songs, and must be able to compose texts extemporaneously. His ability to improvise and ridicule is especially prized, and may account for the close attention people pay to him at funeral performances. During observations of several of these performances, the author was struck by how long people would stand and listen, despite the heat and lack of shade. Each song was sung three times, to constant drumming, and often women would begin to dance. From time to time, a member of the audience would enter the circle and reward a dancer with money.

Many of the songs were described as "teasing," designed to mock, provoke or annoy an "enemy," particularly from another company. They fell into the general category of "songs to reform a wrongdoer." One recalled for the author told the story of a nephew who succeeded his uncle as captain of the company upon the uncle's death. The nephew did not perform his duties well, and the company composed a song pointing out that he was "from the same root" and should therefore be showing the leadership qualities of his predecessor, and living up to expectations. As a result, the person began to take his leadership responsibilities more seriously.

Many of the *asafo* songs would be applicable to circumstances that require energy, determination, and stamina — circumstances that are related to primary health care. Mothers need to be encouraged to bring their children regularly to the baby clinic for weighing and immunizations, despite a long walk in the heat and a busy work day. They need to provide their children with a greater variety of food during the weaning period, and not fall back on the traditional diet of cassava and fish, which although nourishing, is not enough to guarantee adequate growth and resistance to infection and disease. Such songs might also be used in efforts to encourage responsible parenthood, an initiative currently being taken by the Ministry of Health. Efforts to promote family planning are being intensified, the rise in teenage pregnancies is receiving wide media attention, and the issue of child abandonment is being publicly addressed. The engagement of *asafo* companies in the fight for a healthier community seems appropriate to both their traditional and current functions.

Women's *Adziwa* Groups

In most Ghanaian coastal towns, each *asafo* company is complemented by a women's group, called *adziwa* or *adenkum*, the name coming from the songs sung and the kinds of instruments used, usually rattles and gourds. They perform at funerals, "outdoorings" (where a newborn child is brought outside for the first time), and at puberty rites. Each group has a characteristic dance. Dances that are well done are rewarded with gifts of money from the audience.

Songs usually began with a signature tune, which identifies the company to which the group belongs. During the author's observations, the women divided their songs into two categories: songs to reform a wrong-doer, and songs to educate, particularly about marriage. Several songs in the first category stressed that "what you are doing, everybody knows — eyes have seen and ears have heard." Bad behaviours may not always be publicly acknowledged, but they are known. The second group of songs gave advice to young women. One expressed the regret of a young woman who had married against her parents" wishes, and was later maltreated by her husband. Another referred to a plant not able to grow in poor soil, a metaphor for a marriage not able to flourish because the background of the family was not adequately investigated. Yet another called "Why do you sit for your children to cry?" warned young people against becoming parents before they are mature enough for the responsibility. When the children are crying, young parents can't cope. The ability to produce children is not enough — parents must be able to care and provide for them.

Careful attention must be paid, however, to the traditional context of the songs and their adaptability to new ones. At one point during field work, the author suggested that the *adziwa* groups, which often performed at funerals, might address their attention to some of the causes of unnatural or premature deaths (e.g. young mothers who die from neonatal tetanus). But further investigation revealed that funerals are not the setting for anything that reflects negatively on the departed; rather, any expression must be in the form of consolation for the relatives, "to cool the hearts of the bereaved."

Another significant *rite of passage,* however, may hold more promise for change.

Puberty Rites

Puberty rites, known as *nde* in Effutu and *ndako* in Fante, vary from region to region but have similar goals — ushering a child into adulthood, physically, socially, and spiritually. The ritual surrounding the rites is intended to enable the young man or woman "to behave and participate effectively as a useful member of society" (Amponsah 1977: 65). Traditionally, puberty rites were a mechanism for controlling adolescent fertility, and it was a taboo to become pregnant before they had been performed.

However, in the past few years the performance of puberty rites has declined across the country, and Winneba is no exception. The decline is due primarily to the cost of performing the rites, but also to urbanization, the spread of formal education, and to Christianity. Government health workers, traditional leaders, and medical practitioners have all publicly associated the rise in teenage pregnancies with the decline of puberty rites, and have called for their revival and enforcement. Launching a "motherhood campaign" in Ashanti, the district medical officer stressed the need for puberty rites to be used "in a more dynamic and transformed way acceptable to the present generation" ("Use puberty," 1989), as a means of checking teenage pregnancies. Speaking at the launching of the 1990 family planning awareness campaign, he estimated that 78 per cent of females between the ages of 14 and 17 in urban areas are sexually active, yet most of them do not consider contraception ("Legislate minimum," 1990).

The time for the rites in Winneba is usually a few weeks before *Aboakyer,* when the chief or member of his council will beat the gong-gong to inform families with girls of age that it is time to begin. On a given evening, the assembled young women will walk three times around the shrine of *Penky Otu,* the principal fetish of the Effutus, who is considered the custodian and guardian of the Effutu traditional state. They may carry babies on their backs as a sign that they are ready for the responsibility of parenthood. They then return to their father's house for four days where they are "fattened up," and when they emerge, they wear new cloth and jewellery. Their friends and their mother''s friends join in a procession around the town, singing songs that draw the attention of the community, e.g. "The chief's daughter is passing" as well as songs of caution to the girl herself such as "If a man calls you, you shouldn't go." At least one evening of drumming and dancing will be held at her father's house to celebrate her new status, then she will be put on velvet cloth and a black headdress decorated with gold ornaments, and will go around the town to greet all those who participated in the ceremony.

Interviewees in Winneba noted the rites "check immature conceptions" and ensure "moral purity." The textbook for cultural studies currently used in secondary schools (published in 1989) lists the functions puberty rites fulfil:

> They teach the young girl to be obedient to her parents and prevent her from becoming pregnant before the initiation ceremony.
> They enable a girl to acquire some capital for her future, since she receives gifts of money from parents, relatives and friends.
> They help the girl to learn some of her traditions, namely dressing, drumming, and dancing.
> The girl acquires knowledge about "good womanhood."
> A girl who is able to undergo the initiation brings honour to herself, her parents, and her family.

Parents, teachers, and members of the national December 31st Women's Movement have suggested ways to transform the rites, to make them more acceptable to girls and to their families. To cut the cost, schnapps or local gin could be used in the ceremony rather than slaughtering an expensive sheep. Girls could cover their breasts, as those who have been converted to Christianity are embarrassed at exposing themselves. The week-long festivities could be reduced to a day or two, so that girls who are at school would be able to take part. The restoration of the rites, even in an adapted form, could result in a reinforcement of valued traditional norms, and perhaps play a part in checking unwanted pregnancies.

Besides songs and ritual performance, another popular oral communication form, the proverb, was identified in Winneba.

Proverbs

Ong (1982: 35) says proverbs in oral cultures "form the substance of thought itself." In such cultures, restriction of words to sound determine not only modes of expression but also thought processes. You know what you can recall. Devices to aid recall, *aides memoires*, thus become essential.

> Your thought must come into being in heavily rhythmic, balanced patterns, in repetitions or antitheses, in alliterations and assonances, in epithetic and other formulary expressions, in standard thematic settings . . . in proverbs which are constantly heard by everyone so that they come to mind readily and which themselves are patterned for retention and ready recall, or in other mnemonic form. Serious thought is intertwined with memory systems (Ong 1982: 34).

Some of the functions of Ghanaian proverbs have been identified in a current cultural studies text for secondary schools:

> They adorn the speech and make it rich and beautiful.
> They bring out the main point of the matter for clear understanding.
> They make an otherwise long statement short.
> They make listeners pay attention to what is being discussed.
> They educate and teach morals.

Proverbs appear to be widely used and widely understood in Winneba. They form part of everyday speech, and are often used to explain a practice or custom. Several are cited here, along with their interpretation and potential application to health care.

1. "Something may smell nice, but after tasting it you realize it is not good to eat." Appearances can be deceptive, for example a child that is fat is not always a healthy child, and may need a diet supplement.

2. "When two persons sit on a stool, one should not get up without informing the other." "The reason two deer walk together is so one can remove the mote from the other's eye." Both of these stress the need for co-operation, mutual trust, and joint responsibility. They could be applied to a variety of circumstances, including family planning practices and parenting.

3. "If you are ripe for marriage, buy your own plates, don't go borrowing." The emphasis here is on self-sufficiency, and for adequate preparation for such an important step. If you do not have the maturity and resources, you are not yet "ripe."

4. "When a man says he has danced until dawn, ask when he

started." "If a naked person says he will give you his cloth, ask of its name." Both of these encourage a healthy scepticism of smooth talkers; for example, the ubiquitous sellers of ointments and pills at lorry parks and on buses.

5.   "The oracle is always consulted three times." If at first you don't succeed, try again. First-time efforts may not be enough; for example, the first vaccination is of no use unless it is followed up by two more.

In a comprehensive study of proverbs in Akan rhetoric, Yankah (1989) notes that their form and meaning are not fixed; they move with usage. However, there is a control mechanism to ensure that the potential for multiple meanings is not exaggerated:

> The potential for negotiable meaning does not mean anything is acceptable for a proverb's meaning, or that the proverb user is in no way constrained in the specific meaning he associates with the proverb in discourse. Rather, the proverb user is guided in his choice of proverb (and the meaning therein) by his known position or attitude in the discourse interaction, what literal statements precede his citation, or statements uttered after the proverb. These defined attitudes and utterances have to coincide with the intended meaning of the proverb (Yankah 1992: 31).

Yankah's (1992) collection of proverbs that reflect positive attitudes to fertility and fruitfulness also includes those that stress the importance of parental care and nurturing, and the quality of family life. Proverbs such as "polygamy spells poverty" and "if you have five wives, so do you have five tongues" draw attention to the economic realities of large families.

All the communication forms mentioned above rely for their effectiveness on performance and artistic skill to some degree. The more accomplished the drummer, dancer, singer, or reciter of proverbs, the greater the attention and appreciation. What needs to be considered is how these performers' talents could be mobilized to contribute to health education in the community. Such a consideration is addressed in the remainder of this chapter.

Cultural Centres As Catalysts

In declaring the period 1988–97 the World Decade for Cultural Development, UNESCO points out:

The recognition by the international community of the need to place culture at the centre of development is already beginning, though not yet widely, to be reflected in practical terms. . . . In most cases, economic, social, and scientific policies continue to be pursued independently of cultural policies, with scant regard for interrelationships or potential complementarity (UNESCO:18) . . . the process of modernisation takes on its real meaning only if it establishes a new balance between the factors of change and the demands for continuity (UNESCO:20).

Agovi (1989a: 15) draws attention to the situation in Ghana:

There is no way one can continue to pretend that culture and artists have nothing to do with a revolution, structural adjustment, or power to the people. . . . . If the "culture people" don't appreciate these developments or are made to feel peripheral to their realisation, there is very little chance that these efforts can stand the test of time. . . . The same energy and the same active funding which go into political mobilisation and organisation in this country should, if not more, go into the promotion of arts and culture.

The Ghanaian government has been making an effort to respond to Agovi's concerns. In April, 1989, it created a National Commission for Culture as part of efforts to revamp cultural policy. The Commission was to co-ordinate the establishment and functioning of Centres for National Culture in all the regions of the country. These centres replaced the Arts Council of Ghana, a centralized body in Accra that limited its activities to promotion of the arts. At the time of this study, seven of the twelve districts in the Central Region, including Winneba, had established such centrs.

At an orientation workshop for cultural officers held in September, 1989, at Winneba, a representative of the Ministry of Local Government enjoined the audience

to explore the cultural potential of their communities and determine the most effective means of harnessing them to support economic, social and political development.

to consult traditional authorities, community elders, associations and groups to discover "the enormous store of tradition, folklore, and cherished values" that are a part of those communities.

to inspire these people and institutions by making them know they are capable of contributions "to their own upliftment and enrichment of the society," and that their contributions to development can be manifested in works of art, music, drama, poetry, and literature.

The Winneba office had clearly defined policy and programme objectives for 1989 and 1990, as well as specific programme activities.

The cultural officer there, a former teacher, musician, and political activist, saw its purpose as creatively developing and propagating Ghanaian cultural values "as the bedrock of national awareness, identity, and emancipation," and integrating these values into national programmes. Apart from encouraging and promoting traditional art and communication forms (music, dance, drama, poetry, proverbs), the officer and his colleagues were seeking ways "to use the cultural media in raising the level of national awareness among both the literate and illiterate population through educational activities." Such activities included drama presentations focused on a national literacy campaign, puppetry shows, workshops for crafts people, and a masquerade competition on New Year's Day that attracted some 5,000 people. A documentation centre was also being set up where information on local customs regarding marriage, divorce, religious practices, and chieftaincy nominations will be available, as well as data on musical instruments, songs, dance forms and performers.

It would seem that the representatives of various government ministries in Winneba — health, social welfare, community development — could draw on the resources of the centre in planning and implementing their information and education programmes, and that skills and talents would be readily available. *Adziwa* groups often perform at puberty rites, where many of their songs emphasize the responsibilities that must accompany "ripeness," and they might be encouraged to perform at other venues, including at the weekly baby clinic. Although *asafo* performances seem more limited, to festivals and funerals, the men are able to express displeasure, through songs and gesture, with an individual (or group) who has violated community norms. Since public chastisement is a recognized form of disapproval, they might direct their energies against "unexpected marriage" (when an unmarried girl becomes pregnant) or other practices that the traditional council is attempting to reform. For example, attempts are being made to reduce the amount of "head money" required by the bride's family from all prospective husbands, as the amount for literate wives, plus gifts of whisky and cloth, is often exorbitant, and couples feel compelled to start families outside of marriage. *Asafo* members might also be enlisted to participate in ongoing family planning education. According to Huston (1985: 16):

> Courageous leaders could put forth the notion . . . that true manhood resides not in the quantity of offspring produced but in the quality of life that parents provide for their children: proper food, clothing and education. Only men can do this; only men can attempt to redefine society's traditional norms of manhood and parental responsibility.

Until now, however, this opportunity for co-operation between the local performers, the centre, and the ministries has not been realized, largely because of two factors, mistrust and misunderstanding, which are examined next.

Mistrust and Misunderstanding

Perhaps the most immediate of these was a degree of mistrust that lingered between some musicians and performers and any form of cultural bureaucracy. According to the cultural officer in Winneba, some of the people whose talents he would most like to encourage and elicit avoided any association with his office. In this particular case, the officer's background, reputation and personality went a long way towards generating new confidence. However, the concept of community as a unified force pulling together to promote local development often seems to be a romantic one. Midgley (1986: vii) points out that although it is naive to argue that local communities can solve their problems wholly through their own efforts

> . . . it is equally naive to assume that a cosy relationship between the centralized, bureaucratic state and the local community will emerge and that political elites, professionals and administrators will readily agree to the devolution of their authority to ordinary people.

Those who have examined changing state-public relationships in contemporary Africa (Hyden 1986; Rothchild and Chazan 1988) have concluded that citizens often show great reluctance to participate in national government schemes, having been repeatedly disappointed in the past. Governments have often extolled the virtues of self-help but have failed to provide the resources necessary to promote it. Thus "community development" becomes little more than a slogan which brings few tangible benefits.

Another factor hindering the cultural centre's role as catalyst between local groups and health care workers was the tendency among educated people, including those staffing the local government offices, to equate local cultural forms with uncivilized activities, fetishism and paganism. According to the cultural officer, much of the opposition in Winneba district came from "Christian quarters," representatives of the local churches on the Center's planning committee. Such attitudes are not unusual among educated West Africans, and their origins are not hard to find. Agovi (1989b: 5) notes:

Although this country is reasonably rich in the establishment of cultural institutions to meet current cultural exigencies, our educational system since Independence has operated largely outside our cultural assumptions and practices. The educational system continues to alienate our youth, our potential intellectuals and leaders from their sources of cultural energy in our society.

Proselytizing missionaries were the first to ban African religion, art, music and other social activities from the school curriculum, and throughout the history of colonial education in Ghana the cultural gap between those who "know book" and those who are illiterate has grown. The Ministry of Education is, however, making efforts to revive cultural forms in the classroom, and has initiated a cultural studies curriculum and an annual inter-school cultural competition that highlights drumming, dancing, poetry recitals and dramatic presentations. Bells that formerly signalled changes between classes have been replaced by the *atumpan*, or talking drum, which students learn to play.

The attitudes of educated people staffing the government ministries, even at the local level, contribute to the lack of coordination and co-operation between the ministries and the centres. The director of the centres in the Central Region described how the Ministry of Health worked in isolation from his office at Cape Coast, the region's administrative capital. Although story-tellers, a proverb specialist, and a drama and puppetry group are available as a "communications team," they are not called upon to contribute to the planning process when health education campaigns are being discussed. The puppeteers have performed on behalf of the Ghana Commercial Bank, advising people to keep their money in the bank instead of under their pillows. But the only time the Ministry has drawn on the centre's resources was to ask for a dancing or drumming troupe for after-dinner entertainment at a seminar. Thus what Nigerian playwright and novelist Wole Soyinka has called "culture as spectacle" is reinforced, and the lack of fruitful co-operation between institutions continues.

Using Local Resources to Promote Health Care

The roots of resistance to indigenous cultural forms must be acknowledged. Yet traditional media, messages, and settings have been identified that *do* hold potential for communicating health care information and education, and this potential should not be ignored. The concept of health used here is that endorsed by the WHO at Alma-Ata (1978), that health is a state of physical, mental, and social well-being, and that primary health care involves the coordinated efforts of all sectors of society. The WHO initiative recognizes the importance of enhancing and fostering the role of a variety of indigenous resources in health care delivery.

To this end, it has been shown that *adziwa* songs emphasize the responsibilities that accómpany marriage and motherhood, and warn young women against assuming these roles when they are unprepared. There may also be an opportunity in *adziwa* performances to use the cloth the women wear to reinforce their message. Although cloth as a communication form was not part of the scope of this chapter, its use as such has been documented (Sarpong 1974; Idiens and Ponting 1980; Kreutz 1987). The colours, patterns and designs on cloth all carry meaning. For instance, a popular cloth called *Fie woansema* (houseflies) reminds viewers that flies know what is going on in a household, and they carry the news out with them. "If there is any secret about you, someone from your house will know, and they may tell." Such cloth could reinforce messages intended to counter spousal abuse.

Asafo performers are able to express displeasure through song and gesture with an individual or group. Their interpretation of the story of the Akan bird, *Anoma Kokonekone*, who muddies the water as he travels upstream, then complains that the water is dirty when he returns downstream to drink could be directed towards fathers who complain about the cost of large families, but make no effort to limit their size. (Or towards men who complain that their children are often ill, but insist on being served their food first, and with the biggest portions).

In terms of message development, the anagogic framework used by Nichter and Nichter (1986: 65) serves as a guide. In their work in health education in southern India, the Nichters observed that primary health care workers were often "unable to bridge the conceptual gap between two cognitive universes in which they lived and worked." Health and nutrition information was not introduced within a referential framework familiar to villagers, but rather presented in a didactic manner. Such monologues were largely ineffective in introducing new ideas about health care and in addressing people's health concerns.

The Nichters (1986) studied the methods of communicating new information used by popular religious leaders, indigenous medical practitioners, astrologers and politicians. This gave them an appreciation of the effective use these people made of metaphor and analogy, and the enthusiastic response usually accorded them by local audiences. They identified "points of convergence" between indigenous and scientific thinking about health and physical development, and compiled a list of familiar metaphors, analogies, and proverbs. Such an approach, perhaps combined with performance, might be used at the clinic in Winneba, where new mothers bring their babies once a week for immunizations, weighing and nutritional advice. Each week someone from the health team gives a talk on some aspect of child care. One such talk observed during field work was on "drug abuse," that is, buying pills and ointments from unlicensed peddlers and getting shots from local injectionists. The male nurse who *delivered* it gave a straightforward, serious lecture. Observations of the peddlers on buses and at taxi parks revealed a very different form of presentation — they were masters of rhetorical style.

Making use of those already adept in presentation skills would address to some extent Melkote's (1987: 45) observation that "there are hardly any professionals trained to communicate effectively in the peasant idiom." He cites a "pro-literacy" bias in many studies that recommend literacy campaigns as the primary solution to comprehension constraints.

Nevertheless, the limitations of such forms need to be recognized. Starosta (1974) has developed useful criteria for decision-makers, particularly regarding the skill, credibility, and social status of performers. If skilled performers are in short supply, the cost of training becomes a consideration. For most Ghanaians, however, a cultural compatibility with dramatic presentation — what Kidd (1979) has called "a nation-wide aptitude" — may make the form more readily accessible. Certainly the credibility of a performance, and a message, may vary with the source. For instance, certain *adziwa* women may be more acceptable to particular groups than others; those most skilled in the use of proverbs and their pertinent application will gain the most attention.

The concerns that many forms cannot be substantially changed, as their appeal depends upon well-known themes stereotypically presented, also needs to be addressed. In the African context, the practices of improvisation in oral presentation is widely accepted. Finnegan (1970) refers to its "verbal variability," pointing out that stories are constantly transformed by professional narrators, and that there is often no standard form or "authentic" version. Elaboration and extemporization are valued by both performer and audience.

The cross-cultural variations in the acceptability of such forms are perhaps epitomized by analyses from India and the Caribbean. In a conference paper addressing the "possibilities and limits" of traditional, communication forms for development, Thomas (1989: 6) asserts that ritual language, such as that used in Indian folk theatre, is typically "a closed language — it does not allow for alternative meanings." In a paper prepared for the same session, Charles (1989) argues that the view that such language is limited is "misguided," and that what are perceived as limitations are in fact challenges facing the advocates and users of traditional communication forms.

Summary

If cultural officers in Winneba are to be the facilitators between local performers and health educators, they will need to rely on ingenuity, initiative, imagination, and courage — attributes held up to them at their orientation course. To quote one of the speakers:

> This complex array of people, representative institutions and performance outlets needs to be brought together and harmonised for coherent programs. In this regard, the cultural officer has to acquire the flexibility of a public relations manager who is also capable of co-ordinating and reconciling divergencies (Agovi, 1989b: 4).

The enthusiasm and commitment of many of the young men and women present was evident. The considerable challenge facing them lies in confronting the values and attitudes towards traditional forms that have been acquired over generations of colonial education and missionary activity. In order for health care planners and workers to engage local resources, they must learn to recognize the value of traditional idiom and performance. They must also be willing to work with the center and the performers to develop and/or appropriate songs, stories, and proverbs that fit specific instructional objectives, and they must come to appreciate the performers as potential educators. Such a task is a formidable one for all concerned.

NOTES

1. Adekunle's study, carried out in three residential areas of Ibadan, concluded that time, attitude, and ignorance on the part of mothers were three major impediments to child immunization. Two of the responses regarding lack of time ring true for today's circumstances in Ghana:

    > "My children got immunization against smallpox because the man who vaccinated them came to the market. I cannot go to the hospital to vaccinate my children against other diseases because I cannot leave my trade. . . . Please help us tell the government to bring vaccination to us in the market." (Mrs. A., a 38-year-old petty trader, and mother of six)

    > "I did not get my first child immunized because the father did not tell me to do so. But I tried to get the second one immunized and could not complete the dose because we had to go to Abeokuta, my husband"s home town, for his uncle's burial, for six weeks. By the time I came back, I have forgotten about the immunization." (Mrs. P., a 23-year-old woman with two children, one of whom has polio).

Chapter 9

TRAINING AFRICAN MEDIA PERSONNEL: SOME PSYCHO-CULTURAL CONSIDERATIONS*

Louise Bourgault

Introduction

The need to train African media personnel and to foster the growth of media professionalism in African media institutions has been implicit in the move to indigenize African broadcasting and the African press. Despite the enormity and the complexity of this task, media scholars, by and large, have devoted little attention to what such training might involve, either from a pedagogical perspective or from a socio- or psycho-cultural one.

The discussion of media training and of its development in Africa tends to be treated in the literature either descriptively or as a subset of wider ideological issues. Descriptive studies tend to focus largely on the institutions and facilities available for media training.[1] A few studies, however, address the thornier issues such as the local in-service versus overseas formal education debate.[2]

Authors more concerned with the ideological issues bound up in media training have tended to focus on the strains caused by the exportation of Western normative values, including free press and objectivity, to African nations where such practices may clash with local political realities or the local operating definition of "development journalism."[3] Ansah, in Chapter 2, believes the meeting of the two sets of values need not be a "clash" and that the problem is more likely to lie in the inability and unpreparedness of proponents of transplantation of Western communication systems to explore the possibilities of congruence as opposed to eagerness to superimpose technology on tradition. In fact, more globalistic ideological analysts have tended to regard media training (or other western-originating "high tech" training) as tools of cultural and economic imperialism.[4]

While all of these approaches have merit, the majority have a common shortcoming. They fail to address and to adequately discuss basic and crucial issues concerning the human dimensions involved in transfer of technology. In other words, they do not explain the implicit assumptions underlying the

* The editor gratefully acknowledges the cooperation of *Africana Journal* in allowing this article to be reprinted. The original appeared in *Africana Journal*, 1994, vol. 16, pp. 51–65.

belief that training is necessary or why training can be so difficult. They thereby avoid addressing the knotty problems experienced by rural persons from oral societies whose lives have brought them into a head-on collision with technology, and whose work requires them to operate within its logic.

The present study attempts to fill some of the gaps in the literature on media professionalism by taking an approach from the bottom upward. It looks at ordinary individuals whose working lives require them to function within the parameters of modern media organizations in Africa. It examines the work-related behaviors and attitudes mostly of low-level media personnel: camera operators, sound technicians, radio producers, television directors, news reporters, and graphics and set designers.[5]

Most of the observations made in the present work derive from my own experience working with African mass media personnel over a seven year period.[6] It is my contention that many of the behaviors and values I have observed in African media institutions can be properly understood as psychological correlates (psychodynamics) of the oral cultural tradition. But before proceeding along these lines, it is useful to consider briefly the oral tradition in Africa.

The Oral Tradition

From the works of numerous anthropologists and folklorists, we learn of Africa's oral tradition — a great body of largely unwritten poetry and prose. The poetry includes panegyrics, elegies, religious and lyric pieces, hunting and work chants, topical and political songs, and children's songs and rhymes. Prose categories include such diverse forms as myths, stories, legends, drama, proverbs, riddles, oratory, prayers, curses, word play, and verbal formulae. One very special mode of expression is the famed drum language, a complex communication system which encodes messages in stylized drum rhythms for their diffusion throughout communities.

Professor Ruth Finnegan, who has compiled much of the available "oral literature" from Africa, stresses that its major charactertistic is its very orality. Unlike written work, oral literature has no life independent of its performer. The connection between its transmission and its very existence is a highly intimate one. While Finnegan's work does not treat differences in the mentality these oral styles are purported to create, she refers the reader to such authors as Levy-Straus and Goody and Watt (1978).[7]

Finnegan (1970) notes the concurrence within given African societies of both oral and written traditions. She cautions readers against the all too often made assumption that African societies or individuals operate in either the traditional or the modern worlds with no overlap. Societies with both high levels of illiteracy and vast numbers of barely literate school leavers (literate, typically in a foreign language — Arabic, English, or French, for example), are hardly likely to repudiate oral forms of literary expression. Politicians typically rely upon the help of traditional bards to compose political songs designed to reach illiterate masses. And even highly educated university lecturers who entertain socially are not above employing praise singers to "panegyricize" orally the virtues of hosts and guests.

Professor of African literature, E. N. Obeichina, adds to this discussion by suggesting that the oral tradition persists in Africa because: 1) Africans are still largely illiterate, 2) most live in traditional, cuturally and linguistically homogenous village settings which foster oral culture, and that 3) those who do live outside village settings still maintain close contacts with their orally based roots. In sum, he argues, that the oral tradition persists in Africa because it expresses a consciousness which is more typical of the oral than the literary tradition.[8]

The writings of Professor Walter Ong shed considerable light on the nature of this oral consciousness and have thus proved seminal in providing a framework for understanding the African values and behaviors I have encountered in different media training seminars I have given throughout the continent. The central tenet of Walter Ong's work is that the technology of writing structures the consciousness of its users. He believes this to be true on both an individual and on a societal level, though his writing concentrates primarily on the latter.[9]

Ong (1982) operates within the tradition of technological determinism that includes the work of classicists Eric Havelock and Mircea Eliade; psychologists Lev Vygotsky, A. R. Luria, David McClelland, Jack Goody and Ian Watts; linguists Edward Sapir and Benjamin Lee Whorf; political economists Alex Inkeles and David Smith; as well as communication scholars Harold Innis, Marshall McLuhan, and Elizabeth Eisenstein.[10] Ong's (1982) work focusses on the profound effects writing has had on the cultures which have absorbed the technology of writing, transforming them from oral to literate societies. The process is an historical one which began in the Near East with the development of the ideographic writing and continues well into the present day.

To understand the profound changes writing has brought to the oral world, Ong (1982) invites us to consider the dynamic of discourse in a pristine oral culture in a society where "no one has ever looked up anything." In such a culture, he notes, words or facts have "no visual metaphors." They are not symbols. Rather "they are occurrences, events." He adds that oral peoples commonly, and probably universally, consider words to have great power, even magical potency.[11] With no physical repositories for knowledge, persons living in oral cultures must recall or "call up" this knowledge. In these cultures, then, knowledge is stored in memorable ways, that is, in mnemonic patterns shaped for ready oral occurrence. Stored thoughts are recalled from the brain in heavy stylized patterns, alliterative, rhymed, or assonant, in formulary expressions, and in standard thematic settings.[12]

Professor Ong's (1982) work provides a discussion of psychodynamic qualities or mental habits typical of the oral mind set. In contrast with the literate mind, these qualities are: additive rather than subordinative; aggregative rather than analytic; redundant or copious; conservative or traditionalist; close to the human lifeworld; agonistically toned; empathetic and participatory rather than objectively distanced; homeostatic; and situational rather than abstract.

Let us now examine how the dynamics of the oral narrative manifest themselves in African psychology. It will then be possible to understand how the oral mindset is reflected in values and behaviors of African mass media personnel.

Professional Media Behavior and Values in Africa

Ong (1982: 37–38) describes the oral tradition as "additive rather than subordinative." This characteristic is related to the propensity of the oral narrative to use the conjunctive "and" rather than such subordinative expressions as "then," "while," "because," and so on. The use of the subordinate clause requires that the end point of an utterance (or a story) be known in advance. In other words, literate stories stress syntactics, the logical and chronological order of a story. The oral version is focussed outward, on the audience. Because story telling, like music, is participatory[13], the endpoint of a tale is unknown by the griot at the outset of the narrative. Should the audience be particularly responsive, the bard can simply add on details. A writer, in contrast, will add no elements to a literary narrative without planning them in advance. And he or she will not do so unless the additional components advance the storyline. Additive tendencies suggest that the oral world structures time in a way profoundly different from the literate world. Indeed Obeng-Quaidoo (1985) notes that time in Africa "is a two-dimensional phenomenon, with a long past, a present, and virtually no future."

These considerations put into sharp explanatory focus numerous practices I observed in radio and television production in Africa, especially in Nigeria. These are the following:

1. There is a general lack of time orientation in most areas of broadcast production: programmes rarely begin on schedule; programme lengths often exceed allotted time; schedules themselves are often replete with errors; and production crews rarely arrive on time.

2. There is a general resistance to the use of scripts. There is a tendency to allow programmes to run free-form, much like the oral narrative.

It is important that trainers and media managers in Africa understand the oral cultural origins of "untimely" behavior exhibited by studio personnel. This will prevent the former from attributing such tendencies to a lack of staff discipline or to a defiance of authority.

Professor Ong (1982) notes that in the oral tradition, narrative elements are presented in stereotyped clusters or conventionalized constructs. This is a stylistic device which functions to aid recall. In this way soldiers are rendered as brave, maidens as beautiful, monkeys as clever, foxes as wily, and so on. Ong (1982: 38–39) calls this psychodynamic tendency "aggregative rather than analytic." Ong (1982: 42–43) also observes that actions in oral narratives are always attached to activators — to persons, objects, or spirits, and never to forces or abstract constructs; calling this characteristic "close to the human lifeworld." In discussing the phenomenon he adds that there are no lists in the oral narrative, for a list would suggest an abstraction devoid of human action context. The closest thing to a list is a geneology, and these are always presented in the oral world in terms of personal relationships, hence the Genesis formulation of Adam's descendants. Since oral narratives are close to the human lifeworld, oral thought tends to be situational rather than abstract (Ong 1982).

Ong's conclusions draw from the work of Soviet psychologist A. R. Luria (1976) among others. In conjunction with the Soviet collectivization of Uzebekestan farming, Luria studied the cognitive processing of kishlak (village) dwellers in the early 1920s. The literacy levels of Luria's subjects ranged from nil through marginal to fully literate. Luria's results are highly pertinent to this discussion and merit close examination.

Luria (1976) tested his subjects' ability to perform a range of abstract thinking exercises. In one test, he presented geometrical shapes and asked subjects to name them. The illiterates consistently identified these objects by giving them names of familiar objects. Thus, they would call a square a mirror; a circle, a plate; and so on. Full literates used the appropriate (abstract) geometrical term while the marginals responded inconsistently, sometimes providing an abstract name, sometimes an association with a concrete object.

In a second concept formation test, Luria (1976) presented a series of four objects — hammer, saw, log, hatchet — and asked subjects to identify the object which should be eliminated. Fully literates understood the game and selected the log. Illiterates responded to the question by attempting to associate with a concrete and lifelike situation. The following response from a 25 year old illiterate peasant is typical of answers from the unschooled group:

> They're all alike. The saw will saw the log, and the hatchet will chop it into pieces.
> If one of these has to go, I will throw out the hatchet. It doesn't do as good a job as
> the saw (Luria 1976: 56).

The answers of the marginal literates alternated between the two (abstract and situational) response styles.

Luria's (1976) illiterates also resisted giving definitions or providing abstractions. In another test, Luria (1976: 87) asked peasants to explain various objects to persons who had never seen such objects. Situational responses (responses "close to the human lifeworld") were typical "definitions" illiterates would give. Explaining a car, for example, one offered, "If you get in a car and go for a drive, you'll find out." The illiterates also had difficulty making articulate self analysis. When Luria (1976: 26) would ask illiterate peasants to describe themselves, they would refer him to others in the manner of: "What can I tell you about my own heart?" as replied one peasant or "How can I talk about my character, ask others;" or "They can tell you about me. I myself cannot say anything." "Judgment," writes Ong (1982: 55) of the oral world, "bears on the individual from the outside, not from within." Sometimes Luria's (1976: 15) peasants would respond to the self-analysis questions by proffering group norms: "We behave well — if we were bad people, no one would respect us." In other cases, questions designed to elicit self-analysis would generate situational answers: "I came here from Uche-Kurgan, I was very poor, and now I'm married and have children" (Luria 1976: 150).

It is possible to argue that the responses Luria (1976) gathered were poor because illiterate peasants were unfamiliar with the kinds of questions being posed. Ong (1982) addresses this objection succinctly:

> Lack of familiarity is precisely the point! An oral culture simply does not deal in such items as geometrical figures, abstract categorization, logical reasoning processes, definitions, or even comprehensive descriptions, or articulated self-analysis, all of which derive not simply from thought, but from text formed thought. Luria's questions are schoolroom questions associated with the use of texts, and indeed closely resemble or are identical with standard intelligence test questions got up by literates. They are legitimate but they come from a world the oral respondent does not share.

Aggregative thinking can be observed in African media both in the routine production of output and in the values, attitudes, and work orientation of broadcasters. It can also be observed in the discourse style of some African journalists.[14] Television and radio production demand the simultaneous coordination and control of numerous production components. It also demands a linear, or stepwise approach to problem solving. To achieve optimal audio output in a radio or a television production, the audio operator must adjust each individual sound source (mics, music, sound effects, etc.) and then adjust them in combination on the master potentiometer. A voltimeter located on the sound board measures the volume output and indicates when the system is overmodulating. Overmodulation leads to sound distortion and calls for a reduction in volume.

Audio operators I trained in Nigeria tended to concentrate only on the total sound level, not on the individual contributing components.[15] Thus, when the studio VU metre "peaked," for example, these audio operators would respond by lowering the total sound output rather than searching for the one possibly overmodulating sound source. They would often allow one excessively loud actor, sound effect, or musical tape to drown out the others.

Camera operators would resort to the same problem-solving style to adjust focus. If the talent in a studio set moved out of focus, camera operators would focus the camera and vary the focal length (zoom) simultaneously, one behavior often cancelling out the other. I should mention that none of these behaviors is at all atypical of studio novices in any part of the world. I fully expect students unfamiliar with new technology to experience difficuty during an initial phase of operation. What I found startling in the Nigerian context, was the apparent degree with which students resisted a systematic approach to simple studio tasks once such tasks had been introduced, and once the value of these approaches was explained.[16] Globalistic thinking also appreared to govern the entire approach to the studio production process in Nigeria. In the training courses I conducted, I established the use of post-production meetings.[17] I was always surprised at the limited discussion which characterized these meetings. Whether the programme had gone well or badly, crew members seemed at a loss to explain the reasons. It seemed to me that my students had great difficulty isolating the individual components which contributed to the whole production.

Limited abstractive abilities make it difficult for some African media personnel to imagine audiences or empathize with them. In a training course I conducted with Swazi development communicators, I noted the haziness with which these broadcasters appeared to define the characteristics of audiences intended to receive development messages. The following incident illustrates this observation.

In a class on audience targeting, I asked my students to identify and then describe the audience for an intended development message. After much discussion and considerable prodding and hinting on my part, the students were able to draw up, as a group, demographic and psychographic profiles of rural Swazis. (They had already been given a lesson on demographics and psychographics). After a lengthy discussion, one student allowed that he doubted he would be able to change rural behavior. "We can tell them (rural Swazis) what to do, (in this case, to use government-provided pit latrines), but they won't do it." Here, I persisted, stressing the need to learn why rural Swazi homestead dwellers would refuse to use these latrines so that the broadcasters could target messages aimed at eroding the underlying resistence. Another lengthy discussion ensued during which students discussed traditional behaviors and fears related to latrine use. I was surprised at the extent to which students disagreed on the perceived causes of resistance, but this is another matter. What I feel was really noteworthy about this in-class interchange was the novelty of the problem we were tackling. The practice of teasing out human motivations underlying any given behavior was extremely unfamiliar to these Swazi broadcasters. I submit that this very novelty suggests an unfamiliarity with the use of abstractions typical of situational thinkers.

I should like to mention that I have no doubt these same Swazis, operating as rural health workers, one on one or in a small group, could have drawn upon a host of very convincing rhetorical devices to tackle this problem. Indeed, I am fully aware that they command a vast repertoire of persuasive interpersonal tactics. But the effective design of broadcast messages demands a communication competence not necessarily more adept, but of a different sort. It makes requisite a fairly high level of abstractive skills, skills which have had no resonance in the Swazi oral world.

Related to the issue of audience analysis among African broadcasters is the question of self analysis. On this note, I have often been surprised by the limited degree to which media professionals (at various levels) identify with the calling. Elliott and Golding (1979) made similar observations in their study of Nigerian newsmen. Katz and Wedell (1978), quoting from Scotton remind us that recruitment of broadcasters in many Third World countries, most of Africa included, is conducted through routine civil serivce procedures which place little emphasis on the potential candidate's aptitude, interest, or creativity. Broadcasters are defined from the outset as members of the civil service rather than as a special corps of gifted communicators. Hence, to quote Scotton, "it matters little whether they work in broadcasting or in the Ministry of Public Works" (Katz and Wedell 1978: 231).

Civil service careers have slightly less attraction in such comparatively wealthy countries as Nigeria where careers in the private sector can sometimes be more lucrative. Because of this, it seems, many Nigerian broadcasters I trained expressed dissatisfaction with their jobs. It comes as no surprise that such practical attractions as higher salaries or easier opportunities for promotion elsewhere could and did attract individuals to other sectors. What did, however, give me pause, was the imagined (or perhaps unimagined) ease with which many employees expected to transfer professional competencies. When questioned on these matters, many revealed only the vaguest abilities at self analysis or job skill analysis in the abstract. Though many Nigerian broadcasters were dissatisfied with their present positions, only those with concrete practical reasons (low pay, no foreign scholarships, for example) could articulate the source of their dissatisfaction. Many of these individuals would also create elaborate fantasies projecting themselves into, what appeared to me, professions for which they were profoundly unsuited.[18] In contrast with these there were other individuals, equally unsuited to their present media occupation, who seemed content and happy with their performance. Rewards from management, elsewhere for a job well done, moreover, seemed based on criteria other than performance. This management practice undoubtedly did little to spotlight or reinforce professional excellence.

The above points might seem to suggest that media organizations and their employees described herein are unaware of talented broadcasters operating in their midst. But this is not the case. African broadcasters are very alert to their immediate surroundings. An individual who displays talent either before or behind the scenes is eagerly sought after by producers in the organization. But such a person tends to be regarded as more of a fortuitious discovery whose talent should be harnessed in the immediate than someone upon whose skills and abilities a model for recruitment or emulation could be based.

Situational thinking is not only manifest in the expeditious utilization of talented personnel. It also governs the attitude of the African broadcaster toward his job. It seems that media personnel give little thought to their work outside the confines of the station. When the broadcaster or the print journalist leaves the job, he/she enters the world of family, friends, and communal living far removed from the exigencies of media production.

The bifurcation of home life and professional life in Africa leads to many lost opportunites for a professional communicator. There is a general apparent reticence on the part of African media personnnel to engage in work related activity of any kind, off duty, although such "professional" behavior does sometimes occur. The following are examples of tasks often demanded of media professionals outside working hours. They are needed in Africa just as they are needed elsewhere. These include: contacting potential talent (who may be unavailable during the working hours of the communicator, farmers, for example); gathering props unavailable at the studio; conducting "informal" surveys or finding out what the people think about a given issue, in local neighbourhoods or compounds; and even discussing work related problems (for example, my course) outside the organization.

Obeng-Quaidoo (1985: 113) addresses the relationship of the African to work in the so-called modern sector in his discussion of African core values with the following noteworthy comments:

> The African has no full commitment to industrial work since traditional work is more meaningful to him/her than industrial work. For example, farming is a total productive process, whereas industrial work can be short, repetitive, and characterized by unvaried work cycles. The African can respond to firing, layoffs, or unsatisfactory work conditions by returning to the village farm. He/she has in mind that wage is for subsistence living; actual money can be obtained from farming.

> We observe that, unlike industrial man, cultured in the Protestant ethic, and who considers industrial work as a duty connected with one's eternal reward, the African considers work as a necessity for surival, not a duty.

While Obeng-Quaidoo's (1985) point is well taken as regards alienating assembly line work, I believe it is less heuristic in assessing the relationship of a modern professional communicator to his/her rather varied and "people-oriented" occupations. I feel that the concept of "situational thinking" may provide a more satisfactory explanation of the psychological bifurcation of home life and work life by the African media operative.

By situational thinking, I am referring to a propensity I have observed among individuals I trained to maintain professional attitudes and interest at work and then to seemingly shed these qualities at the end of the day. For example, many of the Nigerian broadcasters I worked with were more than willing to expend time and energy (and thought) outside work hours in order to see a pet production project come to fruition. And many often did so despite the nearly insurmountable difficulties imposed by Nigerian urban life. But often as not, radio or TV producers who resolved upon leaving the studio to contact a given talent, or record a specicic musical segment would simply forget to execute the plan. It seemed to me that outside the confines of the workplace, these broadcasters needed cues to trigger their attention to earlier resolutions. Apparently their home environments failed to produce them.

Another incident in Tanzania permitted me to observe and appreciate the concept of situational thinking, in this instance, with a slightly different twist. During a training session I conducted in Dar es Salaam with print journalists, reporters lamented their lack of training on technical matters. "How can we report well on development issues when we don't know the technical aspects of health care or irrigation schemes?" they asked. They followed this highly typical complaint with a standard demand: a call for the government to provide journalists with more formal training. I gave what I now consider my usual response to such comments. I allowed that informal education, education garnered through a curious mind and an eagerness to seize opportunities outside working hours can greatly supplement educational deficiences. I then suggested to this group that they seek out "backgrounders" for development stories from the many technical experts who people the clubs of the Tanzanian capital. These incidentally are the same clubs to which my class of journalists retire in their off hours. My students seemed to find the notion of mixing work and play so novel that they reported my suggestion in their piece about my seminar in the next day's press!

Situational thinking permits a person to behave appropriately in concrete situations but fails to foster in the individual a response repertoire needed for (decontextualized) advanced planning. Situational thinking applied to daily activity also fosters the ritualization of life. Thus certain practices appropriate in any social context are quickly adopted and ritualized while other more individualistic behavior styles are quickly abandoned (Chernoff 1979: 160–1).

The ritualization habit explains another behaviour I initially found curious among Nigerian broadcasters. I noted that they had a strong propensity to routinize apparently random occurrences and to incorporate them into regular practice.

A great deal of studio training in Nigeria occurs on the job, particularly for lower level production personnel. Since relatively little status or importance is attached to these positons, training is done in a very informal manner, often by persons with little more skill than that of the trainee. When I was conducting training at the stations of the Nigerian Television Authority, I often met with resistance when attempting to institute new studio procedures (such as proper camera focussing). The resistance did not surprise me nearly so much as the explanations these junior level studio operatives often gave for their way of working. Attempting to validate a particular (and often incorrect) procedure, young cameramen with as little as two or three months on the job would explain: "This is the way we *always* do it!" as if the lessons of an eight week secondment with another (only slightly more experienced) team member were somehow immutable. Production personnel were equally reticent to engage in crew rotational schemes used typically in American television production classes so as to teach the entire production process to students. Sound technicians had to be coaxed to assume camera positions in crew training sessions, and camera operators were of necessity cajoled into assuming the role of audio operator.

Although this traditionalism was most prevalent among the lower level personnel, it was not limited to them. One general manager, for example invited me to use on-site television facilities for production training courses but mandated that neither the sets nor the props nor the lighting be moved — requirements which quite strongly undermined the course value.

Chernoff (1979: 140) argues that externalized (ritualized) institutional procedures and roles are important to Africans because they "provide a framework and help them to know what is happening and get into it." In other words, the ritualizations and conventions are the means through which the African connects to the larger social world around him.

Ong (1982) would argue that finding one's identity through ritualization is a hallmark of oral culture. Its opposite in the literate is the search for self knowledge through solitary introspection. As I have already noted, solitary introspection together with abstract concept formation increase in society as literacy replaces oral forms of discourse. This process also gives rise to the use of subordinative, time punctuated, and linear cause and effect thinking. And it is by the same means, de-ritualized, exigency-based, and unified "de-situationalized" approaches to problem solving come into being.

Postcript

As African media personnel continue to participate in the kind of training courses presently available to them, they will undoubtedly become more proficient. In so doing, they will become more literate, more skilled at problem solving, and more schooled in machine logic. With these changes, the oral world will gradually recede into the historical past. The nature and the training experienced by these individuals will affect the speed of this process, and the pace at which they in turn transform their audiences.

I believe the process is irreversible. Few, given the option of becoming literate spurn it. Few, given the option of advanced technical training opt to resist it. The oral tradition and the mindset it fosters may be the source of difficulties for mass media trainers and trainees in Africa. Nevertheless, it is with great regret that many witness its passing. I believe the discussion surrounding these phenomena to be of paramount importance.

NOTES

1. Alex Quarmine and Francois Bebey, *Training for Radio and Television in Africa* (Paris, 1967); Mary Katzen, *Mass Communication: Teaching and Studies at Universities* (Paris, 1975); Ram Marathy and Michael Bourgeois, "Training for Rural Broadcasting in Africa," in *Radio Broadcasting Services Development*, No. 48, (Paris, 1965); Frank Barton, *The Press in Africa* (Nairobi, 1966).

2. William Hachten, "The Training of African Journalists," *Gazette* 14, 2, (1968), pp. 101–110; Sydney Head, *Broadcasting in Africa* (Philadelphia, 1974), p.359; Elihu Katz and George Wedell, *Broadcasting in the Third World: Promise and Performance* (London,1978), pp. 102–110; Rita Cruise O'Brien, "Broadcast Professionalism in Senegal," in Frank Okwu Ugboajah, ed. *Mass Communication, Culture, and Society in West Africa* (Munich, 1985), pp. 187–200.

3. Frank Barton, *African Assignment* (Zurich, 1969), pp. 34–54; James Scotton, "Training in Africa," in Sydney Head, ed. *Broadcasting in Africa* (Philadelphia, 1975), pp. 281–290; Katz and Wedell, pp. 102–119; Philip Elliott and Peter Golding, *Making the News* (London, 1979), pp. 114–122; Sharon Murphey and James Scotton, "Dependency & Journalism Education in Africa," *Africana Journal* (forthcoming).

4. Edward Shills, "Demagogues and Cadres in the Political Development of the New States," in Lucien Pye, ed. *Communication and Political Development* (Princeton, N.J.,1963), pp. 64–77; A. E. Gollen, "Foreign Study and Modernization: the Transfer of Technology Through Education," *International Social Science Journal*, 19, 3, (1967), pp. 359–77; O. Stokke, *Reporting Africa* (Uppsala. 1971), p. 94; C. Cooper, "Science, Technology,, and Production in the Undeveloped Countries: an

Introduction," *Journal of Development Studies*, 9, 1, (1972) pp. 1–18; Peter Golding, "Media Professionalism in the Third World: the Transfer of an Ideology," in James Curran, Michael Gurevitch and Janet Woolacott, eds., *Mass Communication and Society* (Beverly Hills and London), pp. 291–308; Herbert Altschull, *Agents of Power* (London, 1984), pp. 169–75.

5. Because of the shortage of educated manpower in Africa, media organizations are often forced to recruit individuals with less than the equivalent of an American high school education to fill these positions. See Elliott and Golding's discussion of the educational levels of Nigerian newsmen, pp. 170–76; See also Louise M. Bourgault & jasmer Singh Narag,"In-Plant Training Sessions Among Nigerian Television Studio Personnel, *Communicatio Socialis Yearbook*, 7, (1988), pp. 160–170.

6. I have spent over seven years working with mass communication organizations in Africa. From 1973 to 1975 I served as a United Nations Volunteer with UNESCO's Educational Television Project in the Cote d'Ivoire. Later I taught practical and theoretical courses in the Department of Mass Communication, Bayero University, Kano, Nigeria, from 1980 to 1982. I then served as a consultant and a media trainer with Media Development Consultants, Ltd., in Kano, Nigeria. In this capacity, I conducted media training workshops for the Federal Corporation of Nigeria, the Nigerian Television Authority, the National Institute of Public Communication, the *New Nigerian* Newspapers, and the Hadeja Ja'amare River Basin Development Authority. I have also run media training workshops in Tunisia (1986), Haiti (1986), Tanzania (1987), Rwanda (1987), Madagascar (1987), and Chad (1987); Central African Republic, Uganda, and Zaire (1990); and Benin and Nigeria (1992) all under the auspices of the United States Information Agency. In 1987 I taught research methods for development communicators in conjunction with U.S. AID and the Academy for Educational Development in Swaziland. In 1988–89 I served as Chief of Party for the USAID-funded Liberian Rural Communications Network. Along with numerous administrative tasks, I worked in manpower training, teaching research methods to LRCN field assistants. Finally in 1992, I worked as a journalism consultant preparing a Project Paper for USAID in conjunction with the development of a Journalism Business Center for private journalists in Rwanda.

7. For a discussion of the relative benefits of the extended case study approach, see Louise M. Bourgault, "Participant Observation and the Study of African Broadcast Organizations," in Sari Thomas, ed. *Language, Thought, Media, and Culture* (Norwood, N.J., 1989), pp. 342–355.

8. Emmanuel Obeichina, "Transition from oral to literary tradition," in *Livre et communication au Nigeria: essai de vue generaliste. Presence Africaine*, pp. 140–161. Another of Obeichina's work examines how the oral tradition is suffused in a popular form of Nigerian market literature known as Chap Books. See Emmanuel Obeichina, *An African Popular Literature. A Study of Onitsha Market Pamphlets* (London, 1973).

9. For a more complete discussion of this theoretical perspective, see Dennis McQuail,

Mass Communication Theory: An Introduction, 2nd ed. (Beverly Hills and London, 1987), pp. 97-99.

10. Eric Havelock, *Preface to Plato* (Cambridge, Mass., 1963); Mircea Eliade, *The Sacred and the Profane: The Nature of Religion*, Translated by Willard R. Trask. (New York, 1959); Lev Semenovitch Vygotsky, *Thought and Language*, Translated by Eugenia Hanfmann and Gertrude Vakar. (Cambridge, Mass., 1962); A.R. Luuria, *Cognitive Development: Its Cultural and Social Foundations*, Translated by Martin Lopez-Morillas and Lynn Solotaroff. (Cambridge, Mass., 1976); David McClelland, *The Achieving Society*, (Princeton, N.J., 1961); Jack Goody, *Communication in Evolution*, (Cambridge, England, 1977); Jack Goody, *The Domestication of the Savage Mind*, (Cambridge, England, 1978); Jack Goody and Ian Watt, "The Consequences of Literacy," in Jack Goody, ed. *Literacy in Traditional Societies*, (Cambridge, England, 1968), pp. 27–84; Edward Sapir, *Language* (New York, 1949); Benjamin Lee Whorf, *Language, Thought, and Reality* (Cambridge, Mass.,1957); Alex Inkeles and David Smith, *Becoming Modern* (Cambridge, Mass., 1974); Harold Innis, *Empire and Communication* (Oxford England, 1950); Harold Innis, *The Bias of Communication* (Toronto, 1951); Marshall McLuhan, *Understanding Media* (London, 1964); Elizabeth Eisenstein, *The Printing Press as an Agent of Change* (2 vols.) (New York, 1978).

11. See also D.I. Nwoga, "The Concept of Satire Among the Igbo," *Conch Magazine*, (1971). Nwoga postulates that fear of satire is at least one of the mechanisms operating against press freedom in Africa.

12. For a socio-esthetic discussion of African drumming, see John Miller Chernoff, *African Rhythms and Sensibility: Aesthetics and Social Action in African Musical Forms* (Chicago, 1979).

14. For a treatment of the oral tradition and its effect on the discourse style of Nigerian newspapers, see Louise M. Bourgault, "Orality and Literacy: An Analysis of the Nigerian Press." *World Communication*, 16, 2, (1987), pp. 211–235. See also, Louise M. Bourgault, "Press Freedom in Africa: A Cultural Analysis," *Journal of Communication Inquiry* (forthcoming)

15. I conducted numerous in-house workshops with television station personnel. These included NTA Kano (Summer and Fall 1981); NTA Maiduguri, (Summer 1983); NTA Bauchi (Fall 1982 and Spring 1983), NTA Jos (Spring 1983). See Bourgault, 1988, pp. 168–170.

16. I do not believe this resistance represented a personnal response to me or my colleague on the part of our students, for these students regularly sought out and followed much of the advice we offered. I should also like to mention that the contrast between these broadcasters (many with as little as eight years of formal education), and the university level students I taught was marked. University level students were well able to follow the demands of machine logic once these lessons were presented in their university-based television production courses.

17. Post production meetings are used in broadcasting for purpose of crew debriefing. During these sessions, individuals discuss the production process they have just undergone. Typically they argue over what went well, what went badly, and why.

18. Painfully honest/shy chaps would imagine future careers as dealers of Nigerian art works/curios, for example. This profession calls for roughly the same personal qualities as are required conventionally for used car salesmen in the United States.

PART III

THE DYNAMICS OF INDIGENOUS COMMUNICATION

Chapter 10

COMMUNICATION IN AN AKAN POLITICAL SYSTEM[1]

Kwasi Ansu-Kyeremeh

In Chapter 2, Ansah dismissed the impression often created that there is no tradition of free expression in Africa. In doing so, he referred to the Akan political system to illustrate some of his points. A description and analysis of a Bono (the Bonos occupy the mid-western part of Ghana, straddling across the border with la Cote d'Ivoire) version of communication within the Akan political system is the focus of this chapter.[2] The objective of the discussion is to show the free flow of information as it was facilitated by various communication mechanisms, structures and institutions in Akan power relations. The discussion begins with the following brief theoretical background.

Theoretical Views

It is generally believed that political factors are important determinants of communication within societies. The MacBride Report which examined the character of information flow between industrialized nations and the developing world in another sense observed that:

> ... communication should be considered a major development resource, a vehicle to ensure real political participation in decision-making ... (Traber and Nordenstreng 1992: 52).

Ito (1990) for his part believes there is an interplay of the elements of government, the press, and public opinion in the formulation of communication policy. Althusser, Gramsci and Freire among others further indicate the desire of the power elite to appropriate the ownership and control of the means of communication to safeguard their interests. The communication patterns in the Akan political system do not reflect such monopolistic and hegemonic tendencies and could therefore have some lessons for the "modern" technology-based media-dominated political systems.

For the purposes of this discussion, political communication is defined as authoritative communication in which the forces that determine the communication process are derived from the political structure. The political structure under examination is dualistic; Ghana being among the "developing countries [in which] many of the traditional forms of communication exist side by side with the modern media" as Bielenstein (1978: 16) observed. There is the Western-style state which is highly centralized with a unitary government. Its communication methods imitate the use of technology-based media systems. At the local level though, indigenous chieftaincy structures exist in juxtaposition with the Western local government structures. It is the indigenous model which is being considered in this discussion. The Bono environment is that of a rural situation where much of the original structures and elements of the indigenous political and communication practices survived the assault of westernization or modernization.

Communication in Akan mainly comprises speaking and listening.[3] To be a good speaker, one must be a good listener. However, reverence and respect are key elements in any exchange of words; and so is a good knowledge of the established lines and patterns of communication. Silence, unless requested by customary practice, such as during the ablution of sacred stools by the Chief (Hogan 1968), or for a particular reason, is considered an insult. This creates a situation in which communication must necessarily be two-way. A well-known local proverb, *Onipa a oretwa sa nnim se n'akyi akyea*, literally the trail blazer is unable to predict the problems that would be left behind without the advice and observations of others, underscores the need for criticism and feedback in policy formulation. Another proverb, *Ti koro nnko agyina* (one mind cannot deliberate communal issues by itself) stresses the need for wider input into political debate. An implication of these notions is that since features such as feedback and two-way flow are essential to the communication process, access to the means of communication must be guaranteed. Consequently, ownership of instrumental channels of communication, some of which are described later, was public and held in trust at the palace on behalf of the community. The exception was the *asafotwene* (*asafo* drum) which was kept with the *Safohene* (commander of the militia).

Even so, a kind of centre-periphery relationship characterized the Akan political structure except that it did not function in the way and manner described by Wallerstein and others. The Wallerstein model resembles an osmotic and parasitic relationship in which the centre draws and feeds on the resources of the periphery. In the Akan model, the relationship between the centre and periphery (units) is more symbiotic with both centre and periphery feeding on each other. Unlike the centrifugal political organization of the Western political models in which the sources of power and communication radiate from the centre, the Bono political structure in its purity was centripetal — the centre derived its power from the co-operation of the peripheral units. Communities at the geographical periphery of the political system were by no means marginalized in the political communication, the reason being that communication was an integral part of the social interactive patterns within the polity.

Communication was more inclusive, participatory, and universally accessible in its indigenous pre-European form. Communication was also holistic; an integral element in a total social system which could not be isolated from the other elements of the system for any meaningful examination. In contrast, current geographically defined peripheral status translates into marginalization in the communication flow within the contemporary Western-oriented political system. Authoritative communication has become absolute control of Western-oriented communication systems by the dominant political group thereby restricting, or even excluding, the rural periphery from ownership, input into, or access to the means of political communication. Communication in this model tends to be more unidirectional and hegemonic.

The political significance of communication may be defined in terms of freedom of communication and freedom to receive information. In other words, ownership, and particularly access to the means of communication, and also to the content of communication, are often considered indicators of the extent of free flow of political ideas, beliefs and positions within a society. For those, such as Althusser, who believe the means of communication as an apparatus the dominant power in the society controls for the purposes of perpetuating itself, the state's attitude toward the media is important in measuring the degree of liberty the society is likely to experience. Evidence on indigenous communication systems, however, tend to suggest that they are not the kind of media that render themselves to easy monopoly, control and manipulation by a small section of the society. It is helpful here, then, to examine the role of the type of communication they generate in the political process.

Dynamics of Political Communication

Four basic actor elements are prominent in the Bono political communication system. These are the leader (Chief or Queen Mother), the *okyeame* (linguist or spokesperson), the *dawubofo* (gongman), and the *amanfo* (ordinary people). The Chief or Queen Mother is a principal source and initiator of communication. However, it is also true as observed by Doob (1961: 280–1) that:

the prestige of established leaders is carefully maintained so that in their normal roles as communicators their communications are likely to evoke approval.

Perhaps it is the need to sustain the prestige of the political leadership that prescribe mandatory "personal behaviours" a chief-designate has to be taught during the forty-day confinement period of education prior to assumption of office. These included "talking in public" as well as communication with the spirits of the dead (Obeng 1988: 43–4). Such special instruction also includes nonverbal communication like nodding the head.

Being a principal destination and receiver of messages means a good Chief must have the skill and capacity to listen. Thus, the chief-elect is also taught simple listening skills such as looking into people's eyes as they speak. It is imperative that the Chief knows when to close the mouth and open the ears. In fact, in court deliberations and indeed, in all discussions at assemblies or court gatherings, the last word goes to the Chief or Queen Mother. Where both are present the former speaks last. This arrangement is to afford them the opportunity to listen to as many of their *amanfo* (ordinary people or citizens) as possible to help him/her to develop the capability of reaching sound and plausible decisions.

This communicative method in a decision-making procedures is not unlike the one adopted by the cabinet in a "modern" Canadian democracy. Former Canadian prime minister Pierre Trudeau (1993: 96) noted that the prime minister would "listen as others [ministers] had their say . . . [and] respond at the end voicing conclusions aloud." The process especially enables the Chief to tap the wisdom of his people. It also underscores the inclusiveness of the communication act and the fact that feedback is of utmost importance.

Communication usually begins with the statement of its purpose, for example, an announcement of an agenda, followed by the contributions of all involved. At the second stage, everyone, irrespective of status, is availed the opportunity to initiate communication or access it. Obeng (1988: 30) observed that:

> At a meeting of the village Elders every young man has a right to be present and speak. If the young men in the village as a whole express an opinion on a matter, the sub-chief is obliged to present their views at a meeting of the Division.

Each member of the community is encouraged to play the dual role of a sender and receiver in the communication act.

Next to the Chief, and probably of greater significance in Akan political communication is the *okyeame*, also called *linguist* in local slang. The Queen Mother or *Obaapanin* (literally female elder or female head), though, "had her own 'Spokeswoman' and her court was constituted of both men and women" (Obeng (1988: 13–14). In a two-step communication, messages flowing from initiators of political communication such as the Chief or Queen Mother or even an Elder playing a political role, as well as messages addressed to them, are channelled through the *okyeame*. Yankah (1995) has articulated the *okyeame* as the centrepiece of Akan political oratory.

What an *okyeame* usually does, though, is to restate the message verbatim or paraphrase it. That unique position makes him or her the key instrument of the content filtration mechanism for encoding and decoding messages. Understanding instantaneous "editing" of messages originating from and directed at an indigenous political authority such as the Chief, however, does not make the *okyeame*'s role comparable to the modern newspaper, radio, and television (especially in pre-recorded programmes and prepared speeches) gatekeeper or editor. With the communication transaction process contemporaneous and speeches and addresses extemporaneous, the *okyeame* cannot enjoy the benefit and luxury of time created by the delay allowed editors in the technology-generated communication process. The *okyeame* may consequently not have the accompanying power to slant messages and experience the constraints of its self-censorship implications. This is not to obscure the fact that the *okyeame* is an expert of spin, being

> a man [person] of great eloquence and wit who would remove the sting from a gaffe made by a chief [or queen mother], or embroider his [or her] statement in a pleasing and entertaining manner (Anon. A human encyclopedia, *West Africa*, 1–7 November, 1993, p.1935).

In other roles, the *okyeame* is more than a conduit pipe. For example, s/he acts as:

> a court official who normally attends to the chief and is given a variety of tasks. He is sometimes an intermediary through whom a request is put to the chief; . . . an envoy sent to the government or other courts. He even has his role in judicial proceedings, in view of the fact that records of proceedings do not exist (Anon. A human encyclopedia, *West Africa*, 1–7 November, 1993, p.1935).

The role of a filter to messages originating from, or addressed to, the

Chief certainly controls and moderates mistakes and errors of judgement of what turns out to be the Chief's final pronouncement (as stated by the *okyeame*) and improves the richness of the language while enhancing audition. Again, it provides redundancy at the same time as it controls it through the use of figures of speech, proverbs and synonyms or, as Riley puts it, the local idiom. The *okyeame*, however, is "in no sense an interpreter" (Maxwell 1928: 38).

The role of the *dawubofo* (gongman), as explained in some parts of Nigeria by Ugboajah (1979) and Wilson (earlier in this book), and as with the Western "town-crier", is that of a messenger. Most of the time, he is engaged in one-way communication involving the announcement of a ceremony, the observation of a religious rite, or summoning a meeting. He beats his *dawuro* (gong) to attract the attention of people and delivers the message as soon as a crowd assembles. One of the shortcomings of an announcement carried by the modern broadcast media is the inability to repeat, or provide an opportunity for another look at the content of messages. With the gongman, the audience are at liberty to seek clarifications of the his messages to which he is obliged to respond.

In some cases, the *okyerema* (drummer), or even the hornblower (using the *abentia* or horn), may be substituted for the gongman or may work in conjunction with him, depending on the purpose and what is to be communicated. For example, to cover greater distances or to address those who understand special drum or horn language, the latter may be more suitable channels than the gong. Wilson in Chapter 3 describes various types of the gong, drum and horn instruments some of which have Bono equivalents.

The *amanfo* (the ordinary people) are more than the "subjects" Obeng (1988) portrays them to be. They are citizens who are not passive consumers of information "transmitted" by the decision and opinion leaders. *Amanfo* always seek, and in fact, do enthusiastically exercise the right to respond to any messages they receive and therefore actively participate in the communication process. This is probably the most distinguishing feature of the indigenous communication patterns compared to the Western model of political communication *delivered* through the one-way technology-based media systems. The communicator cannot expect his or her message to be received without any reaction. In fact, one might say that lack of reaction to a message can only be interpreted as rejection or indifference. Silence can never be construed as consent or agreement. Sustaining this process of political communication is a political organization which is described next.

Political Organization

Like the political systems of the many post-colonial societies in the developing world, a dualistic political organizational structure constituted by an indigenous (in this case Akan Bono) and Western political organizational structures operate *pari passu*. Either model affects communication in its own way. The indigenous political sphere is represented in a Traditional Council which exercises authority over a defined geographical Traditional Area and is operated by a well-researched indigenous centralized governmental system.[4] A more detailed description of the Akan indigenous political system (as represented by the best documented, the Asante Kingdom) is undertaken by Obeng (1988). A brief description of the Bono model is provided here to help illustrate its communicative aspects which were described above.

The symbol of power and authority of all categories of Chiefs was, and still is, the *Stool*, the equivalent of the Western notion of a throne. A Stool (a *Skin* in the northern parts of the country) represent the spirit and body of the people and it bestows upon its occupant, the Chief, the privileges, power, and authority to govern. The sight of a stool or skin thus symbolizes power and authority. Succession to the Stool has always been hereditary.

The flat-topped pyramidal power structure is headed by the *Omanhene* (Paramount Chief) of the *oman* (State or Traditional Area). The Traditional Council is composed of the Omanhene (the President), the *Ohemaa* (Queen Mother literally female chief), and nine *Abrempon* ("wing" or "sub-chiefs"), who could also be *asafohene* (military commanders of their wing units), because their titles evolved from military roles in the past. Each of the nine wing chiefs heads an *omansin*, a town, with its own junior *adikro* or heads of the villages which are affiliated to that town. Leadership titles and stature often coincide with settlement size.

All chiefs, from *Omanhene* to *adikro* (headmen) make decisions in Council. Decisions which affect the "state" are, therefore, made by representatives of all sections of the community through consensus, and not by the Chief as an individual. Status aside, any *Ohene* (Chief) or *Ohemaa* (Queen Mother) can summon citizens, within an area of jurisdiction, to appear before his or her Council of Elders (*Mpayimfo*) in the area, for a message to be "presented" (*de asem to mo anim* or literally put a message before you for discussion) or might ask messages to be "delivered" (*ma mo aso ate* or literally let you know). "Presentation" and "delivery" have special implications for communication processes. Presentation is usually to an assembly of people with provision for questions and feedback. Delivery didactic communication transactions involve messages such as directives, which are passed on through the gongman or court messengers, as well as information or reply conveyed through the *okyeame* (spokesperson). The *okyeame*'s role in the communication process, as explained earlier, is to repeat the Chief's words, sometimes by paraphrasing in a kind of redundancy for emphasis and clarity. This is so even during communication by the Chief with the gods and the spirits of the dead ancestors. According to Obeng (1988), the purpose of this communication with the departed is to invoke the blessings of those elements for the living members of the community. Messengers carry an *akofena* (sword) as a symbol of authority; and the *okyeame* carries an *akyeampoma* (staff of office) for the same purpose. With the ownership of all the devices of communication and symbols of authority public and accessible to all, there can hardly be any monopoly of the means of communication under this political system in the manner Althusser has expressed with the Western systems.

Communication and information which flow within this centripetal indigenous political system worked so well in the past that the British colonial administrators described it as "a democratic government to a degree of which there is not any modern parallel in Europe" (Maxwell 1928: 34). Roberts (1976) also observed that the aggregate units of the periphery, the villages, were autonomous in their internal politics. A combination of the political and social institutional structures and processes nurtured the indigenous communication patterns.

Colonial extraneous intervention has altered the character of existing communication patterns. In fact, the relatively democratic political structure is now largely rivalled and dominated by a Western type centrifugal political organization system. The Western system comprises a hierarchy of political units from local through regional to national governments. Regional is subservient and answers to the national and the local is accountable to the regional.

The superimposed hierarchically-ordered Western political model allows minimal policy formulation at the grassroots level. The role of the local level is actually to implement decisions made at the national or regional level of authority. The Western-style structure is supported by a centralized media system which reflects the centrifugal nature of the authority structure. This arrangement is in contrast with the guaranteed input of village communities into decisions made by the *Omanhene* — directly and through their *adikro* or *ahenfo* representatives. Under the Western system, the communities are reduced to "passive" information consumers who live with decisions made by bureaucrats, technocrats, and politicians far removed from the village environment. Often, problems of the "communication gap" between the centre and the periphery are exacerbated by authoritarian regimes at the national level. Between 1966 and 1981, the country fell under four different military dictatorships, the last of which has been in power since December 1981 (having transformed itself into an "elected" government from January 7, 1993).

Suffice it to point out at this juncture that the Bono socio-political system is holistic with all institutions intertwined. It has therefore not been a practice to compartmentalize the Akan social structure into institutions such as education, politics, and the like and deal with them individually. The communication formats described above were thus inextricably linked to the affairs of the society not directly relating to the politics of the place as discussed next. To shed light on indigenous political communication, the intricate but intertwined social relationships and interactions will be briefly discussed.

Social Relationships and the Indigenous Communication Infrastructure

Social relationships in the Bono villages Ansu-Kyeremeh (1989) studied were largely based on the kin concept, although no single religious faith predominated as is often the case with such communities which also resemble Tonnies hypothetical gemeinschaft. Villagers interacted on an individual basis, or in groups — both primary (family, and marriage) and secondary (church, sports, and social).

Ansu-Kyeremeh (1989) also found that villagers belonged to both primary and such secondary groups. The latter could be classified as social groups (based on sports and fraternities) and religious or church-based associations. There were also in existence secondary groups in the form of political organizations such as a Mobilization Squad, a Democratic Youth League, and a Committee for the Defence of the Revolution. Riley earlier made reference to their branches she observed in a coastal district as "revolutionary organs." Their group-to-group communication with the "non-revolutionary" groups was anything but guaranteed. In the first place, the revolutionary groups exhibited the communication characteristic of that developed between them and their parental organizations at the centre. It was unidirectional accountability communication which did not encourage interaction between the group and its non-governmental counterparts. Second, these were branches of an oppressive military government propaganda and law and order enforcement machine which was required to *disseminate* information and not to share or discuss it. Third, they were quasi-intelligence organizations for whom secrecy was paramount. Fourth, not many people would want to be associated with these political groups. In fact, members of the groups were often treated as *paraiah*. The people's desire to distance themselves from these groups was displayed by an elderly woman who pointed at them as the "government's people." Actually, the fear generated by these truly Athusserian organs of state was so intense that it generated a "culture of silence" within the nation.

As already noted, at official assemblies, the Chief communicates through the *Okyeame*, the spokesperson; but speaks directly to everyone in unofficial interaction. Ansu-Kyeremeh (1989) recounts how during his fieldwork he once bumped into a Chief in conversation with friends the former was visiting. Spouses spoke to each other freely, but each needed to choose words carefully when communicating with in-laws. An *Okomfo* (priest of an indigenous religion) and Elders were greeted (especially by women), with a bow or a genuflection; and men would slip their cloth off the shoulders in deference. Thus, excepting the Chief-subjects official communication, no rigid communication barriers, arising out of the above social differentiations, were observed. Spouses, parents and offspring, people of different religious faiths, males and females, and adults and children shared free, two-way communication with one another.[5]

Intra-Village Communication Patterns

That communication is the essence of village community was borne out by the intricate but clearly defined network of intra-village communication patterns that Ansu-Kyeremeh (1989) observed among the villagers he studied. Nine different types of indigenous and more than four Western technology-based communication systems were identified by him. The indigenous ones could be distinguished from each other on the basis of source of message likely to be channelled through that specific medium, the nature of the message, the target group for the message, the basic purpose of communication served by that medium, and the process and direction of the particular communication. On this basis four basic classifications could be made. They are: venue-oriented communication, events as modes of communication, the communicative nature of games, and performance-oriented games. Ansu-Kyeremeh (1989) found high participation and access rates among villagers in these indigenous communication systems which are described next.

Venue-Oriented Communication

One pattern of village communication was manifested in venue-based communication transactions. In this mode, certain venues in the villages served as meeting points where conversations naturally developed. At these gatherings, information was passed through structured or unstructured conversation among people present.

Venues where this pattern of communication was generated were identified. They included the Chief's or the Queen Mother's court and verandahs. Others were artisans' (blacksmith, carpenter, potter) workshops, *nkwankwaannuase*, *gyeduase* (tree shade), *kookoo aboe* (cocoa pod splitting venue), *nsadwase* (drinking place), *ahoroe* (riverside washing sites), and *afukwanso* (trails leading to farms).[6] *Nkwankwaannuase* is so called because it was a popular gathering spot for the youth (*nkwankwaa*) who together constitute a formidable political force and have their own chief, the *Nkwankwaahene*.[7] The *Nkwankwaahene* will always be consulted by the *Ohene* [Chief] in any thing which will involve the young men . . . (Obeng 1988: 30).

Meetings at these spots were informal and in small groups with the presence of people purely on a voluntary basis. Ansu-Kyeremeh (1989) cites examples of these gatherings on the senior *ahenkwaa*'s verandah; at a carpenter's shop; in front of the house of a village Elder where people had gathered to play the *oware* game; young women's hairdressing; young boys' meeting at the house of a teacher; and at blacksmiths' shops in various villages. A small village market was another type of venue. In addition Ansu-Kyeremeh (1988) believes that even the confines of a moving vehicle plying between villages served a communication purpose.

At these venues, people were involved in interactive conversation. Conversations centred around practical and useful information or focused on the importance of early vaccination of children. Many learned of goings-on in the villages at these venues. The meetings could be in large groups (sometimes involving the entire village adult population, especially when there was a formal summons or invitation).

Village assemblies or rallies may generally be summoned on behalf of a government official or a stranger by the Chief's authority (exercised in council) at any of these venues where convenient. In the past the information officer, the community development officer, and health education teams took advantage of this setup.

Events as Communication Modes

Institutionalized events (usually celebrations made up of a series of ceremonies) also facilitate communication of information. Ansu-Kyeremeh (1989) observed that the annual *Kwafie afahye* (festival) and *bragro* (a girls' *rites de passage*) are two such events. They were occasions when villagers gathered together to participate in various rites over a period of time. They were particularly suited to information campaigns. *Kwafie* ceremonies had a very high participation rate.

Games as Communication

Ansu-Kyeremeh (1988) further observed games like *oware* and *dame* (draught) which were played in some villages facilitated communication and information exchange. Various political topics and issues were often discussed in the course of the games.

Other games he observed which provided opportunities for discussion were *aso* (locally called *teere*), and *ampe*, both all-girl games, which combine song and movement of hands and feet. Into the lyrics of the songs would be formulated meaning of political significance such as recording, describing, or recalling events and episodes that occurred in the surroundings and praising or ridiculing people. It was a powerful means for sanctions in the form of stigmatization of individuals who by their actions and social behaviour had violated some custom or tradition. Heroic acts or exemplary behaviour, however, earned commendation. *Ampe*, a synchronized rhythm of song and movement of feet and hands (clapping) was less serious in its messages but nonetheless an important means of communication.

Performance-Oriented Communication

Also observed by Ansu-Kyeremeh (1989) in the Bono countryside were drama and other forms of song and dance which displayed great communication attributes. Drama was organized within the village or by extra-village commercial *concert* groups, of the type described by Bame (1985), which visited the villages to put up paid performances.

Other expressive and performance-oriented communication systems identified among the villagers were the *adowa* (Marie identifies it as *adziwa* in another part of the country) drumming, song and dance, *nwomkro* (hymns) and *abee* (funeral dirges) songs, and puppet shows. *Nwomkro* and *abee* involved the singing of songs, the latter mainly funeral dirges. Performers were usually all adult females. *Adowa* was a combination of song and drum language with both male (on drums) and female (singers) artistes. In *adowa*, dancing (by both men and women) was an essential response to the drumming and songs which were couched in to convey messages in a manner not unlike the Hawaiian *hula* (Anon. 1994: 52). Riley earlier attributed communication functions to the *adziwa* which is a southern variety of the *adowa*.

Narratives such as *anansesem* (storytelling), and speech forms such as *anwonsem* (poems), proverbs, puns, rhymes, and runes were other useful communication forms. The educational capacity of *anansesem* is widely acknowledged especially when it comes to the transmission of mores and values. Puppetry and a brass band were also spotted.

More Actors in Communication

The principal sources of the message in the just described communication systems were largely the decision leaders in the villages. As already explained, the decision leadership included the Chiefs, Queen Mothers, Chiefs-in-Council, and Elders whose titles, status and roles were ascribed. They were message originators themselves or they would play a surrogate role (in a two-step delivery format) to the Western-style political authority by relaying externally-originated authoritative messages (information, orders, or directives) which were channelled through them.

Other message sources or initiators of communication were the opinion leaders, whose status and roles were achieved and who included teachers, artisans, drivers, and visiting urban-resident relatives. Students in higher institutions were particularly an exceptionally powerful force. Others in the opinion leader category were leaders of performing groups (artistes such as the lead singer or drummer), parents, peer group leaders, and hawkers. Prominent members of the village community with higher achieved social status such as the herbalist or craftsperson, and who commanded social prestige and some form of authority, also belonged to the opinion leader category. An *okomfo* (priest) or a *dunsini* (herbalist) would always speak through an *okyeame* although in this instance the latter is more of a translator (especially when the communicator speaks in tongues). Messages originating from these sources often targeted villagers in general, or discriminated on the basis of gender or age.

One notices that virtually all the communication systems (except drum, or other instrumental, language) are orally-based and interpersonal.

Ansu-Kyeremeh (1989) observed that the flow of messages channelled through them effectively combined vertical top-down and bottom-up flow with horizontal flow. They were usually two-way and dialogical. For example, *mmoguo* (interjection) facilitated listeners' participation in the *anansesem* (storytelling) process. And in court deliberations questions were permitted. In court communication transactions, the Chief was the last to speak after drawing on input from all present and trying to forge a consensus from divergent views and ideas. The only exception to the two-way rule would be when in a two-step flow, a government agent visiting the village passed on a message over which he or she had no control, or when instructions were didactically passed on. In whatever form, and for whatever reason, communication, especially when it originated from the political leadership, was initiated or facilitated by signals transmitted through a number of instruments of the like in Wilson's inventory. Some of the most prominent ones are briefly discussed next.

Instruments of Communication

A number of instruments, drums, horns, and the metal gong provided communication signals. The *akyerema* (drummers), *dawubofo* (gongman), and hornblowers as agents of communication operated by drums, the *dawuro* (metal gong) and *abentia* (horn) respectively. These agents or messengers operated the instruments on behalf of the political leadership of *ahenfo* (chiefs), *ahemaa* (queen mothers), and *mpayimfo* (elders). Even the *akomfo* (priests) sometimes speak through such instruments especially the drum. Drum instruments, of the membrophonic type described by Wilson (Chapter 3), include the *atumpan* (talking drums), *twenesini* (beaten for emergency village assembly meetings), and *asafotwene* (beaten as clarion call to the *asafo* militia to summon members to emergencies such as search and rescue operations). The *dawuro*, which is identical to the metal gong *nkwong* in Wilson's inventory, is the main instrument of·the *dawubofo*, the gongman who Wilson calls the official "broadcaster." The *abentia* is the Bono equivalent of the *nnuk enin* of the Ibibios of Nigeria as described by Wilson in Chapter 3. It is an ivory horn manufactured from an elephant tusk and held only at the Chief's court.

Wilson indicated that the sound signals of these types of instruments were sometimes coded and could only be decoded by a section of the population. In the Bono villages, although this was true in some complex drum languages, there were some basic signals that were universally understood. Members of an *asafo* might be the only ones to understand what exactly their *asafotwene* would be saying. However, to the non-*asafo* person the mere sound of that drum, which was so scarce that sometimes it would not be heard in a generation, always meant trouble. Every villager was aware that he or she was required to head for the Chief's court at the sound of the *twenesini* even though the *mpayimfo* who were more educated in its language would be the only ones to decode the nature of summons for the meeting. The *dawuro* has an even more familiar signification, so simple that some parents would send their children to listen to the message.

Generally, then, the indigenous communication patterns within the village communities were characterized by the spoken word and interpersonal face-to-face interaction. They often involved an audience of two or more and encouraged a feedback process. Credibility was often assured, because instantaneous verification of facts through questions, queries, contributions and comments was possible. This was demonstrated whenever participation and feedback was encouraged. It has already been explained that existing *pari passu* with these communication patterns in the Bono polity were some Western technology-based mass media types such as radio and radio-cassette recorders. There is far more literature available on the operation of the latter media systems within the environments of the like of the Bono political and social milieu. What this chapter has sought to do for the unfamiliar who have to operate in cultures with characteristics similar to the Bono is to propose consideration of the issues raised regarding the interface between the indigenous forms of political communication and the pursuit of development objectives.

Summary

When it comes to the utilization of indigenous communication systems, as they affect political and social interaction among the Bonos, especially in the rural areas, the economic climate may not be of great importance because, as Ugboajah (1985) noted, these forms of communication are accessible without any financial costs to the audience. Immediacy of access and time-saving are guaranteed as are important elements in the communication process such as feedback and audience participation. The indigenous communication systems are derived from, and integral to the socio-political structures of the Bono communities.

What development planners and agents who operate in those circumstances need to know is that, properly accessed, the indigenous communication systems can be of immense assistance in communicating effectively. But if one is to avoid sending any signals of disrespect or disdain that will poison the political atmosphere, there is always a need to undertake a survey of the communication patterns to establish the directions of flow, the actors and the procedural etiquette. As Doob (1961: 281) noted, even with the local decision leaders such as Chiefs who enjoy a lot of credibility and authority, they put all that on the line when they

> pursue Western ways and are unable to acquire the proper symbols, their prestige as communicators is low and their communications may be ineffective.

Some patterns of flow and actors have been identified here. They are likely to apply in most Akan situations and among many other rural African communities. It is up to those who find an understanding of the political communication process useful to delve further into its mechanism to achieve the maximum effectiveness in communicating in environments such as the Bono polity.

NOTES

1. Partly based on observations from field research for 1989 doctoral thesis, *Communication in Rural Education: An Exploratory Study of the Application of Media to Ghanaian Village Education*, Centre for the Study of Educational Communication and Media, School of Education, La Trobe University.

2. The most famous Akan political system is the Asante model which has been discussed several times over. One of the most recent descriptions is Ernest E. Obeng's *Ancient Ashanti Chieftaincy*, Tema, Ghana Publishing Corporation, 1988.

3. Doob (1961) observed that there is also a significant nonverbal component which operates under various codes of conduct. Examples include signs, symbols, gesticulations and general body language.

4. The 1928 edition of the annual Handbook edited by J. Maxwell provides detailed description.

5. While according to Babacar Sine (*Non-formal education and education policy in Ghana and Senegal*, Paris, Unesco, 1979, p.11) "As regards social relations clear distinction is drawn between men and women and also between the young and the old," gender and age are no serious communication barriers. For example. K. Agbetiafa (in Gaston Mialaret, *Out-of-school education*, Paris, Unesco, 1979. pp. 195–202) dismissed the often held characterization of the "child is seen and not heard" as mythical).

6. *Gyeduase* is the shade created by the canopy of a big tree, especially soothing on

hot afternoons, where people gather for conversations or games. It may also be known as *nkwankwaannuase* where the gathering is mainly by the youth.

7. Their urban counterparts known as the *verandah boys* (and later described as the *sans cullotes*) during the independence struggle were very active in political parties.

Chapter 11

"NI NDE?": A CASE STUDY OF RADIO SOAP OPERAS IN BURUNDI*

Devote Ngabirano

Soap operas are an illustration of "entertainment-education" or "edutainment" in mass media. The strategy of "entertainment-education" appears to be a successful means of contributing to social change. One may also remember the success of *telenovellas* in Central and South America, cartoons in many parts of the world, and soap operas in the Caribbean Islands and in India. Burundi, a small country in Central-Eastern Africa can also be added to the list of countries which have grasped the importance of education through entertainment as illustrated in its radio and television show *"Ni Nde?"* (which means "Who Is It?").

This chapter intends to demonstrate that although *"Ni Nde?"* differs from other soap operas in that it lacks commercial purposes, the show is a potential powerful means to deliver messages of social change. The principal motivation in selecting this subject is to bring attention to this kind of programming in Burundi. In addition, since the creation of *Ni Nde?*, no one has evaluated the show to see whether it has met the objectives that were assigned to it, or what the audience feels and thinks about it. Therefore, the chapter will also attempt in a small way, to contribute to the study of soap operas in general, and Burundi in particular, in terms of radio programming.

Developed from an exploratory study which is designed to stimulate more research, the findings presented here are mainly descriptive. They are based on responses to face-to-face interviews and a questionnaire instrument administered to fourteen Burundians (one woman and thirteen men) who were all living in the United States at the time of the interview. What is recorded here are their expressed opinions about the show as an entertaining-educational tool in social change, and what future orientation they would like the show to take.

The chapter first traces the historical background of *Ni Nde?* and its beginnings to see what factors contributed to its development. Secondly, the show is described in terms of its themes, plot, setting and characterization. The third section makes a synthesis of the participants' opinions on the

*An edited version of paper of the same title presented at the Institute for Advanced Study and Research in African Humanities. Northwestern University April 9th–10th, 1993.

show; what other people known to them thought about it; and suggestions concerning any changes that they would like to see in the future episodes of *Ni Nde?*.

Beginnings of *Ni Nde?*

"*Ni Nde?*" emerged from conditions that were both internal and external to the Burundian socio-cultural system. Internally, one notices that Burundi has strong oral traditions which have existed for a long time. In addition, the political willingness to revive and encourage aspects of the Burundian culture, to avoid dependency on foreign programs, motivated the creation of locally-produced programs in Kirundi (the country's national mother tongue).

In fact, it is not surprising to find in Burundi the "news" method of delivering messages for social change. Beyond that situation, one can also note that the concept of entertainment-education is not an invention of "modern" thought in communication. It can be traced back to old civilizations in many parts of the world before the repeated disturbance of what is considered to be Western civilization. The contribution of "modern" theories of communication seems to be the conceptualization of the term. Furthermore, entertainment-education is found in many traditions with oral literature (folk tales, folk games and songs, rituals and ceremonies). Numerous examples can be found in the indigenous communication systems identified by Wilson, Riley and Ansu-Kyeremeh in the earlier chapters of this book. In African societies, that traditional element managed to survive the European destruction of various elements of the indigenous cultures. Even today, a quick glance at "global" African music reveals that descendants of Africa in the Americas and other parts of the world, who have been cut off from the continent for centuries, still deliver revolutionary messages and cry for justice and respect for human rights through song and dance. In the tradition of the strong Burundian oral traditions, elders, wise women and men used to teach younger generations by means of entertainment-education.

With the disruption of the Burundian society and the introduction of Western education, some of the country's previous ways of educating its children faded away but did not disappear completely. The Western school made it difficult to keep the tradition of the "evening classes" when members of the family, spreading out to include the extended family, would interact through different forms of entertainment-education. However, it takes time to obliterate strong customs and traditions that are deeply rooted in the lifestyle of a given society. As is shown in this chapter later on, *"Ni Nde?"* is one such tradition that has refused to disappear.

It has been difficult to establish the exact year *"Ni Nde?"* began, especially from the distance the author finds herself, from where she is unable to access official documents. However, conversations with interviewees traced its beginnings to the early 1980s. Participants in the study were equally unable to state the exact year of its beginning. To the question "When did *'Ni Nde?'* start?" all the participants hesitated or replied that they had no idea, either because they were outside the country, or did not notice. Answers included "Around 1982/83 on radio, and 1985 on TV;" "Maybe 1984 or sometime before;" "Probably 1984/85;" "It started drawing my attention in 1985/86." Even a former employee of the Burundi radio station who worked there as a broadcaster for about ten years, and was director of the station before he became *Secretaire National Charge de l'Information, la Formation et la Mobilisation* of the UPRONA Party, was unable to remember the exact year. However, he situated the beginning of the show in "about 1983–84." According to him, the project could be traced back to a long time before that date. In his own words: "When I started working at the radio station, about 1979, the project, it seemed to me, was already there. And it was sometime later [1983–84] that . . . the first episodes were produced."

The ex-broadcaster went on to explain the immediate context of the creation of the show. He described the circumstances so accurately that I prefer to report his account (translated from French) in its entirety:

> Indeed, how was it born? There had been a provincial competition (between the different provinces of the country) and in the competition, one had to detect the different talents that existed in the different provinces on an artistic level and elsewhere. And there happened to be a group which was dynamic enough in the region of Gitega, and something was tried and it ended up working. There came an idea that one could train the group and have them play in radio drama (pieces radiophoniques), and this produced the first episodes/series of *"Ni Nde?"*

This quotation highlights three significant elements that determine the nature of the show itself. First, it is a show created from *within* and not from the outside which makes it an example of emphasizing endogeneity in African communication systems as stressed by Epskamp.

Although one cannot deny inspiration from "radio plays" produced outside Burundi, the fact of looking for internal talents all over the country reflects a willingness to start something authentic. Secondly, the account illustrates another aspect of contemporary theories of mass communication, the participation at all levels of the society, instead of the "hypodermic" or one way model of communication that was popular in the 1950s and 1960s. Thirdly, there is a recognition of innate creative and artistic talents of rural Burundian people themselves, and the possibility to combine those talents with modern communication technologies such as the electronic media.

It appears then, that the provincial competitions to locate gifted people in dramatic performance illustrated an aspect of the political orientation of the country's leaders in the early 1980s. Indeed, although this paper lacks excerpts of records of the political speeches of the time, it was a period of the Second Republic when the regime contemplated a "cultural revolution" using the media as instruments. Many new shows in Kirundi were created; there was a renewal of interest in folk songs, participation of the elder generation (who one might consider the nation's "library") in radio programs. *"Ni Nde?"* became part of the long term programs through which it was hoped Burundian culture would find a counter-balance to the cultural neo-colonialism the country experiences through the electronic media, and reinforce Burundian identity.

Part of the willingness to preserve national heritage was the need to educate citizens about themselves. Although some participants in this research did not know the reason why the show was created, many of them gave either what they heard or what they "speculated" to be the motives behind the introduction of the show. The responses sometimes completed or contradicted each other. For instance, a doctoral student in environmental science (at the time of the research) who was also a former teacher and dean at the *Institut Pedagogique* in Bujumbura, stated that the motivation was "to educate people, provide them with an education based on the traditional values of the country." To a military student at the USA Army War College, there was an educational purpose behind; "the proof," he said, "is that the show is in the country's native language so that a lot of people listen to it." A graduate student in Business Administration also saw in the show a need "to educate the masses."

It is interesting to analyze what the participants meant or understood by "educational." From their different responses, two opposing views emerged. For one of them, "education" meant a return to traditional values that the country might lose unless it made an effort to preserve them. For another, to "educate the masses" was to facilitate a "change of mentality." Both people may mean the same thing, but the most common and biased understanding of the change of mentality when talking about non-Western societies is to abandon one's old cultural values in favour of "new ones" (generally Western), unless one speaks like Ngugi Wa Thiong'o in terms of "decolonizing our minds." Both participants also seemed to agree with a third whose view was that the show "intends to correct" the negative side of the society. It seems then that the participants, by their comments, believed some form of learning was taking place.

According to the participants, another motivation in creating the show was the promotion of art and entertainment. For instance, one called "*Ni Nde?*" a comedy which was created "probably to utilize actors' and actresses' talents to produce programs in the national language." He added that the show would also "entertain people who listen to the radio, by offering them a unique program." A graduate student in broadcast journalism, who was also a former radio broadcaster in Burundi, compared "*Ni Nde?*" to works by famous seventeenth-century French writers, who were basically critics of their society during their time. "Taking into account the content of the program," he said, "one finds a little bit of Moliere and Beamarchais, a little bit of La Fontaine in his fables." Consequently, he saw "*Ni Nde?*" as a satiric show, since these French artists delivered their socio-political message through satire. Whether the show was consciously conceived to fulfil this role or not is difficult to tell. But the participants' opinions show that "*Ni Nde?*" combines elements of entertainment and education. This observation leads to the form and content of the show which are discussed in the following section.

Form and Content of "*Ni Nde?*"

Until the summer of 1992, "*Ni Nde?*" was being broadcast on the radio on Sunday at 8:30 pm for an hour, and on television Tuesdays and Fridays at 7:30 pm (also for an hour). As mentioned earlier, there has not been any survey to assess the audience response to the show. Nevertheless, all participants in this project were unanimous about its popularity. One indicator cited was the fact that on viewers' requests, "*Ni Nde?*" was being broadcast on television twice a week although it had begun itself as a weekly program. The popularity of the show is discussed in greater detail later in this chapter.

For now, though, it is the subject matter or content of "*Ni Nde?*" which is explored. When asked about what they thought "*Ni Nde?*" dealt with, the participants answered with references to descriptions of the plot and the usual setting, and also pointed out numerous themes. Respondents to both the interviews and the questionnaire recognized that the setting was most of the time the Burundian rural area. Even when episodes were set in the capital city (Bujumbura), the focus was on the behaviour and problems of new migrants who came from the rural areas.

The plot depended on the topic of the episode and, apparently, it followed the traditional story-line of exposition, development of conflict, climax and resolution (that comes with "catharsis"). This last part usually came with an implicity epilogue and message which suggested a way of solving the problem addressed, or a lesson to learn from the experience. The stories developed numerous and varied themes including the rural outmigration exodus, family and community relations, extra-marital affairs, marriage and dowry and wife abuse by irresponsible husbands. Other themes centred on adolescent delinquency, greed, witchcraft, farm life, mismanagement of public funds and corruption by civil servants. Alcoholism, patriotism, betrayal and dishonesty, and other social issues were the themes at other times.

A close look at the themes shows that "*Ni Nde?*" deals with issues on three levels which are connected to each other: individuals, family and community, and national. At the individual level, the show seems to promote "moral" values such as courage, honesty and determination; and discourages vices, undesirable behaviour caused by envy and jealousy, ignorance, as well as "immorality." At the family and community level, the show explores difficulties related to marriage, the treatment of women, behaviours of wife, husband and children, and some customs and practices that need to be reexamined.

At the national level "*Ni Nde?*" depicts problems that the Burundian youth face especially when they leave the countryside thinking that the city would offer a "better" life to them. The problem of migration to the city is a real problem in many, if not all African countries in the contemporary period, and African writers have also addressed the issue. Other themes involve land conflicts, the mismanagement of family welfare and the generation gap. According to a respondent[1], these themes are addressed to both people living in the countryside and in the city. Moreover, "Ni Nde?" has melodramatic elements in its "thematics" which include "conflict of kinship, . . . parenthood, the struggle for family survival."[2]

The participants further discussed the characters developed in "*Ni Nde?*" When it comes to human qualities of values and vices, respondents recognized that the characters embody human feelings and emotions. As far as relevance was concerned, nine of the participants said that the characters "reflect a lot of what's going on in the country." The characterization exposed, among other things, the selfish and extravagant husband, the administrator and other political authorities under him, the civil servant, parents and relatives, the neighbour (friendly or unfriendly), the sorcerer, the "medicine" person or indigenous healer versus the "modern" doctor or physician, the judge and many more. In his comment a respondent elaborated on how the characters reflected Burundian daily life:

> These characters are so realistic that the listener or viewers can replace any actor/ actress by someone he/she knows in real life. The show reflected well people I knew, men, women, children and legendary people.

Another added, "They were not my twins . . . but they described a microcosm that was also mine." This particular comment seemed to suggest that the characters depicted in the show are realistic and believable. "*Ni Nde?*" actually reminds one of *The Archers* on the BBC in that it deliberately provides educational information about farming and other issues, via an entertaining storyline.[3]

Another important aspect of "*Ni Nde?*" concerns the script and the writer. "*Ni Nde?*" has no single script writers. A participant who knew about the conditions of production of the show said, "Ideas come from everywhere." Sometimes, someone from the team would generate an idea, present it to the performers and ask them to develop and play it. The show is thus more of a team work; and it preempts ownership of the script by one writer. Besides, some of the actors are illiterate.

One respondent observed:

> It's amazing how people who don't know how to read and write can hold roles and play every time. . . .You have the impression that they have something like a script, but they don't.

The absence of a written script does not prevent the actors of "*Ni Nde?*" from performing well (according to most respondents). The technique of unwritten script is partially and successfully used in the Ghanaian popular television drama series *Osofo Dadzie*. The story about "*Ni Nde?*" including its format, its themes and the characters it deals with can be further understood by exploring the theoretical framework of entertainment-education.

Theoretical Framework of "*Ni Nde?*"

Although there is no evidence that "*Ni Nde?*" is a conscious application of theories of mass communication, the show contains elements of social learning and communication theory in many ways. Social learning theory posits that humans learn social behaviours as a result of modelling their behaviour on that of others with whom they interact or whom they observe.

Thus, individuals observe and imitate behaviours of other real life or fictional people, based on rewards and/or punishments they see others receive for their actions.[4] These assumptions find illustration in *"Ni Nde?"* by the way in which characters are depicted, and issues are addressed. Positive role models get reward at the end of the story, while negative ones are punished or killed off. For instance, one of the respondents talked about an episode of *"Ni Nde?"* which compared "thatched huts" with "modern" housing architecture. The character who refused to adopt the new way of building (the "laggard" in Everett Rogers' diffusion of innovation model) ends up having his house on fire (the idea being that grass catches fire very easily), while the "adopter" lives a happy life.

In addition to role modelling, *"Ni Nde?"* is founded on identification by creating relatively believable characters, and selecting topics that Burundian people can identify with. Comments from respondents about their degree of identification varied from "that's the way life is in the countryside;" to "I never experienced that in my family;" or "I could identify with what some characters were going through;" and "I knew a young man who lived the same experience as the central character in the episode." The responses are contradictory and prevent anyone from drawing a definite generalization.

"Ni Nde?" also illustrates the theoretical framework for selective exposure, which advocates that individuals pick and choose role models with whom they identify. This selection probably explains the variety of the responses regarding the participants' identification with characters in the show. One respondent related the popularity of the show to this identification with what he calls "everyday people:"

> It's something [widely] experienced. It reflects topics of everyday life, and people like that . . . they recognize themselves in the show. Even the title itself is a question . . . and you try to find out who it is.

According to most respondents, characters in the show vary from the responsible loving mother to the abusive husband; the tolerant, patient and progressive thinker to the impatient radical rebel. This structure of the plot and characterization implies that listeners or viewers are offered choices of behaviour so that they can select who or what to identify with. Essentially then, although there was no conscious theoretical formulation for the concept of the show before it started, one notices that *"Ni Nde?"* is comparable to many other soap operas around the world in terms of its form, content and format. However, unlike the *telenovellas*, for example, the show is longer than most of the well-known soap operas.

Ansu-Kyeremeh (1989) observed that the flow of messages channelled through them effectively combined vertical top-down and bottom-up flow with horizontal flow. They were usually two-way and dialogical. For example, *mmoguo* (interjection) facilitated listeners' participation in the *anansesem* (storytelling) process. And in court deliberations questions were permitted. In court communication transactions, the Chief was the last to speak after drawing on input from all present and trying to forge a consensus from divergent views and ideas. The only exception to the two-way rule would be when in a two-step flow, a government agent visiting the village passed on a message over which he or she had no control, or when instructions were didactically passed on. In whatever form, and for whatever reason, communication, especially when it originated from the political leadership, was initiated or facilitated by signals transmitted through a number of instruments of the like in Wilson's inventory. Some of the most prominent ones are briefly discussed next.

Instruments of Communication

A number of instruments, drums, horns, and the metal gong provided communication signals. The *akyerema* (drummers), *dawubofo* (gongman), and hornblowers as agents of communication operated by drums, the *dawuro* (metal gong) and *abentia* (horn) respectively. These agents or messengers operated the instruments on behalf of the political leadership of *ahenfo* (chiefs), *ahemaa* (queen mothers), and *mpayimfo* (elders). Even the *akomfo* (priests) sometimes speak through such instruments especially the drum. Drum instruments, of the membrophonic type described by Wilson (Chapter 3), include the *atumpan* (talking drums), *twenesini* (beaten for emergency village assembly meetings), and *asafotwene* (beaten as clarion call to the *asafo* militia to summon members to emergencies such as search and rescue operations). The *dawuro*, which is identical to the metal gong *nkwong* in Wilson's inventory, is the main instrument of the *dawubofo*, the gongman who Wilson calls the official "broadcaster." The *abentia* is the Bono equivalent of the *nnuk enin* of the Ibibios of Nigeria as described by Wilson in Chapter 3. It is an ivory horn manufactured from an elephant tusk and held only at the Chief's court.

Wilson indicated that the sound signals of these types of instruments were sometimes coded and could only be decoded by a section of the population. In the Bono villages, although this was true in some complex drum languages, there were some basic signals that were universally understood. Members of an *asafo* might be the only ones to understand what exactly their *asafotwene* would be saying. However, to the non-*asafo* person the mere sound of that drum, which was so scarce that sometimes it would not be heard in a generation, always meant trouble. Every villager was aware that he or she was required to head for the Chief's court at the sound of the *twenesini* even though the *mpayimfo* who were more educated in its language would be the only ones to decode the nature of summons for the meeting. The *dawuro* has an even more familiar signification, so simple that some parents would send their children to listen to the message.

Generally, then, the indigenous communication patterns within the village communities were characterized by the spoken word and interpersonal face-to-face interaction. They often involved an audience of two or more and encouraged a feedback process. Credibility was often assured, because instantaneous verification of facts through questions, queries, contributions and comments was possible. This was demonstrated whenever participation and feedback was encouraged. It has already been explained that existing *pari passu* with these communication patterns in the Bono polity were some Western technology-based mass media types such as radio and radio-cassette recorders. There is far more literature available on the operation of the latter media systems within the environments of the like of the Bono political and social milieu. What this chapter has sought to do for the unfamiliar who have to operate in cultures with characteristics similar to the Bono is to propose consideration of the issues raised regarding the interface between the indigenous forms of political communication and the pursuit of development objectives.

Summary

When it comes to the utilization of indigenous communication systems, as they affect political and social interaction among the Bonos, especially in the rural areas, the economic climate may not be of great importance because, as Ugboajah (1985) noted, these forms of communication are accessible without any financial costs to the audience. Immediacy of access and time-saving are guaranteed as are important elements in the communication process such as feedback and audience participation. The indigenous communication systems are derived from, and integral to the socio-political structures of the Bono communities.

What development planners and agents who operate in those circumstances need to know is that, properly accessed, the indigenous communication systems can be of immense assistance in communicating effectively. But if one is to avoid sending any signals of disrespect or disdain of Burundi; although such an assertion is not without some controversy. Finally, the respondents' opinions about the show express different feelings about its messages. But one thing they all agree upon is that the show be more inclusive in the future, and more varied in its content. Two respondents said that they were bored by the "same kind" of quarrels. They appreciated the creativity in *"Ni Nde?"* but they also added that the show could do better. Some respondents suggested a kind of "professional" training, perhaps of the type Bourgault describes (in Chapter 9), in order to upgrade performance.

The principal merit of *"Ni Nde?"* at this stage of this study is that it illustrates a global African trend, reminding one of productions such as the *Cock Crow at Dawn* television series in Nigeria, and some works of Elaine Perkinds in Jamaica. Yet, the soap opera maintains some unique characteristics being deeply enmeshed in the Burundian experience. *"Ni Nde?"* is also an example of efforts to achieve self-reliance in programming at a time when efforts to improve communication in many African nations direct their energy and resources toward perpetuating their dependency relations with the "North" by acquiring more of Western-derived media products without undertaking the extensive adjustment needed to make them appropriate for local conditions.

Concerning the present research, I need to emphasize that the findings are anything but conclusive about the nature and role of *"Ni Nde?"* and the audience response to it. However, this chapter offers some pointers to future research. For example, a content analysis to ascertain a possible role for *"Ni Nde?"* in fighting burning issues such as AIDS will be most worthwhile. Investigative approaches (including semiotic, psychoanalytic, feminist, Marxist and audience response) can also be employed to analyze different aspects of the show. Who benefits from the show and to what extent? Is the cost of *"Ni Nde?"* worth it? These are legitimate questions for future research the answers to which could yield invaluable information. All one can say for the time being is that *"Ni Nde?"* is an illustration of radio and television programming which is inspired by national culture, and which promotes change to some extent. The show still faces serious challenges in issues of representation, and the wider challenge of improving dialogue and production of material for "global Africa."

Editor's Note

Unfortunately, we are unable to know the final outcome of Devote Ngabirano's valuable research. She passed away even before this book went to press. Let us hope someone will take up from where she left off to provide this important information on the indigenous aspects of communication in Africa. The real names of individuals who participated in the study were cited in the original paper that was submitted. However, they were edited out because of the current political situation in the country.

NOTES

1. At the time of the interview, he was Director of Information at the Ministry of External Relations and Co-operation.

2. Elizabeth Lozano, "The Forces of Myth on Popular Narratives: The Case of Melodramatic Serials." *Communication Theory*. Vol. 2., No. 3, (August 1992), p.209.

3. Arvind Singhal (in press), *The Entertainment-Education Communication Strategy for Social Change*. New Delhi: Sage Publications, p.146.

4. Synthesis of the theory as formulated by Albert Bandura, *Social Learning T¹*. Englewood Cliffs, N.J.: Prentice-Hall, 1977.

Chapter 12

TEXTS IN OBJECTS: THE GENERATION OF A GENDERED TEXT IN MITYANA (UGANDA) WOMEN'S CLUB FESTIVAL SONGS AND PERFORMANCES[1]

Helen Nabasuta Mugambi

Women's clubs, not unlike the Ebre Society of southeastern Nigerian women and the women's *adziwa* group of the Akan of Ghana discussed by Wilson (Chapter 3) and Riley respectively, have flourished in many rural communities in Buganda region of Uganda since the 1950s. They evolved as a response to the gender inequalities propagated by parents' preference for the education of boys over that of girls following the introduction of the British schooling system. Other women's clubs were being organized by religious sisters who possessed domestic science expertise although the clubs retained a secular status in their dedication to reinforcing leadership in the adult female community.

Under the British schooling system, many women in Buganda were denied the opportunity to acquire reading and writing skills at a time when literacy became a key determinant of social and economic mobility. The consequent marginalization of orality disempowered women who were substantially excluded from the new education system.

The original objective of the clubs was to help women acquire basic literacy skills to enable them to read about aspects of homemaking, particularly nutrition and child care. In addition, club members expanded their skills in areas such as weaving, embroidering, baking and other homemaking skills. Among the crafts they produced were items such as woven floor and table mats, kettle covers, baskets and bags. They made ovens from local materials in order to produce home-made bread and economize their use of firewood. Such a focus on the woman's domestic responsibilities gave the clubs the reputation of being the propagators and guardians of traditional cultural values affecting household gender roles.

In his discussion of music in community life in Africa, ethno-musicologist Nketia (1966: 13) explains how

> the general pattern of music organization is one that emphasizes the integration of music with other activities, with social and political action or with those activities in which African societies express or consolidate their interpersonal relationships, beliefs and attitudes to life.

The activities of the club women reflected this tradition as they used song to accompany or to punctuate weaving, sewing and other club activities. Excelling in song, dance and performance slowly became an integral part of club activities resulting in inter-club competitions. Over time, the women transformed club houses into powerful sites not only for the production of household objects but also for the production of verbal arts and performances. In the fifties and sixties, women's clubs such as those in Naggalama, mainly performed already established traditional songs and dances. It is very clear today, however, that with changing social, political, and educational systems, women are composing their own songs and choreographing dance movements to express themes of protest or national identity. Many of the songs embody social and political critiques of patriarchal ideas inherent in the existing social system. Paradoxically then, the marginalization of orality and women was turned on its head as women reinvigorated their power through song and performance within a space authorized for their further domestication. In other words, the imperialist idea of development indirectly facilitated the formation of women's groups which today marginalize writing as they measure excellence through oral performance for change in gender and national politics.

Gender politics was, from the outset, an integral part of club activities. In order to join a club, a woman had to obtain permission from her husband. Some men would not give such permission as they looked at women's clubs with suspicion fearing that the gathering of women outside the home would create sites for "gossip" and have the potential for changing women's attitudes towards traditional ascribed male authority within the household. On the other hand, those men who allowed their wives to join the clubs perceived the clubs as safe non-threatening domestic spaces that helped to produce skilled homemakers. The findings of a study of recent songs and performances of Mityana women's clubs by the author give validity to both views in that these clubs produce more effective homemakers while simultaneously providing a forum for changing key traditional attitudes about gender hierarchies and responsibilities. It is clear from studying the festival objects and activities that club members gain better homemaking expertise in rural domestic technology. Most importantly, the clubs also served as catalysts for women's articulation of gender relations through song and performance. Although these clubs were not at the outset referred to as development organizations, their activities show that they are actually the precursors of today's development organizations. Today women in these clubs articulate their own indigenous ideas of self improvement or *kwesitula* women's *adziwa* groups in Ghana as they appear in Riley's analysis in Chapter 8. These texts and meanings embody a dialogic relationship with the Kiganda traditional narratives whose texts prescribe woman's status in the Kiganda society.

The relationship between the objects in the exhibits and the texts dramatized on the stage was played out inside a large circular area at the festival site. Exhibit booths lined the bottom three quarters of the festival site while the performance stage occupied the top quarter of the arena. The main entrance led to the central space which was reserved for the people who came to see the exhibits on display and to attend the performances as audience. The movements of the members of the audience, thus, formed a bridge linking the flow of ideas emanating from the activities at the two sites.

There were approximately eighteen display booths designed to exhibit the artifacts and natural objects described above. Although the functions of familiar objects in the booths were not visibly inscribed on the objects themselves, they existed within the cultural consciousness of the audience. Spokeswomen at each booth created verbal texts as they recited or narrated these functions to guests and visitors who came to observe these objects on display. The Mityana women also created texts about their newly invented objects. The overall intention of the women was to present a comprehensive picture of both the natural and woman-made objects essential to the construction of the ideal homestead.

The following extract from a statement by a Kiganda club lady showing me around her club's booth exemplifies such narrations:

> We DEVISED ways of constructing this (decorative) kettle and these cups. . . . These here are Ziba yams; these are Kiganda yams; this is a bunch of bananas . . .; these are beans;... Over there we have Kiganda medicine. This HERB here is known as Kiyondo. It *cures* morning sickness in pregnant women; you CRUSH AND MIX it with water and have her drink it.. . . . This one here is called *bbombo*. It cures chest pains. Now this little HERB is called *mubiri*. It *stops* vomiting in a child. . . . This over here is called Luwawu. It is a POISON ANTIDOTE. . . . [emphasis mine.]

In the display booths, spokeswomen drew spectators' attention to the types and functions of handicrafts, the various types of foods that go into the making of a balanced diet or demonstrated how the energy-saving ovens they invented functioned.

The songs at the performance stage, like the narrations at the booths, named and described the functions of the exhibits and in the process, reproduced and expanded on those narrations. Additionally, the text of the (literally meaning "lifting themselves up") as they compose songs explaining their attempts to transform rural technology and improve their physical and cultural environment. They also articulate their function and roles within the wider socio-political structure in contemporary Uganda.

The Mityana clubs sometimes make their crafts and compose songs and plays around a pre-selected theme. During a three-year period that ended in July 1992, the women's activities focused on home improvement, emphasizing the construction of an ideal homestead. Aspects of the programme included naming the clearing, manicuring the compound, and planting flowers. Other aspects involved upgrading and maintenance of the kitchen and the main house, and constructing dish racks, rubbish pits and animal houses. Also included in the activities was the provision of safe drinking water as well as the maintenance of kitchen gardens. Simultaneously, women were determined to start income generating businesses which included the selling of their handicrafts.

July 1992 marked the climactic celebration of the achievements of the Mityana women's clubs three-year phase. The women came together in a weekend camp to celebrate their domestic and communal achievements. The celebration took the form of an exhibition of handicrafts including handmade mats, baskets, clay pots, and table-cloths. Also exhibited were foodstuffs from their gardens and "state of the art" rural technology. The technological achievements included environmentally-sensitive energy-saving devices as well as home-grown traditional medicinal herbs. The exhibition was accompanied by song, dance and drama. Each club constructed a miniature version of an ideal homestead complete with houses, furnishings, gardens, and in some cases chickens, goats and pigs.

As part of my on-going research on gendered texts in Kiganda songs, I recorded and studied songs and performances composed by the Mityana. Some of these were staged at their July 1992 festival. In this discussion, I examine the relationship between material objects and the oral texts and performances created by the women for this festival. I demonstrate that there exist significant meanings attached to these material objects. These meanings are not visibly inscribed on the objects themselves but are, instead, articulated and contained in the texts of songs, plays and dances composed and performed by the women. Furthermore, I show that in composing these texts the women reproduce a feminist consciousness[2] generating new texts and meanings within an emergent culture of female liberation parrallel to one articulated by Mbulelo Mzamane (1992) in his discussion of gender politics in South Africa. The practice of formulating certain meanings into texts to achieve specific goals is also similar to the performances of the

the booths. This song, like many others performed throughout the day, stressed the necessity to combine household responsibilities with money-generating activities such as selling their handicrafts. This approach to the hard work necessary to the running of a household, they asserted, would result into women achieving economic independence from their husbands and achieving overall self-reliance.

The tune and tempo of the above song abruptly changed with the following text:

(Lady) When you have done all this work
Lady when you have organized everything
Wherever you go,
What do you become?

Light !
    CHORUS

        We are the light
        We are the brightness . . .
        We thank the Creator
        For having made us female

Another song by another group of women utilized the objects in the exhibition to claim that it is woman's labor and creativity that brings health and development to the household. They described a bachelor's household depicting it as devoid of any civilizing grace. They further stated, "*Omukyala Agulumizibwe*" (may woman be exalted); (Curiously, the expression "*agulumizibwe*" which means "exalted" is most commonly used in reference to deities and saints.) This song also stated:

The ladies of Vvumba possess excellence
They excel in all their work
They invent
They are excellent PROFESSORS

In such songs, women reconfigure traditional texts that envisage domestic work as oppressive. They initiate new texts that transform their work into a liberating force. Thus they accept, then transcend oppression in domestic chores. It is important to view these new texts against traditional cultural texts on domestic work. It is only then that the transformatory power of such texts is highlighted.

songs extricated the meaning conveyed by the objects. Such meaning was neither visibly inscribed on the objects nor embedded in the cultural consciousness of the audience. Songs often outlined the significance and functions of the exhibits. A particular song entitled: "*Entereeza Yamaka Nga Byweyandibadde,*" or "How the home should be organized" explained the symbolism of the handicrafts. In that song, the women state in part:

Don't ignore making mats and baskets
They make the home aesthetically pleasing
They make the home shine. . . .

Lady, don't ignore my words
Agriculture is the foundation of the nation
Grow maize, grow beans
Grow food for a balanced diet. . . .

Another song "*Ffe Bannakazadde Beggwanga,*" or "We the Mothers of the Nation" presented the most comprehensive description and narration with the women on the stage acting out the spacial arrangements of each section of the homestead. The women started by declaring their intention to prescribe and demonstrate the layout and content of an ideal contemporary homestead. The song was addressed to "fellow women" and recounted the capabilities and relevance of practically every natural and constructed object on display.

In many of the songs, the women projected themselves as inventors citing their knowledge and expertise in crafting the objects in the exhibits such as energy-saving devices. They referred to themselves as knowledgeable "physicians" in control of traditional medical practices and of curative traditional herbal medicines.

When these songs are comprehensively analyzed, it becomes clear that these women utilized their overall achievements and demonstrated skills in the construction and invention of household objects to create song texts through which they established or legitimized female power and authority within the household. Hence the texts the women attached to the objects generated yet other texts, texts about female power, about female identity and about women defining themselves. This process, indeed, moved the centre of discourse from the objects on display to gender relations in the household. The above-quoted song entitled "*Entereeza Y'amaka*" by Busubizi women prescribes woman's responsibility in the household exhorting her to hard work including the making of crafts as displayed in will govern the household. The man orders his wife to clear the bush, plant food and cash crops and give any money generated from the work to him. Her husband squanders all their money on alcohol and expects her to perform all household duties singlehandedly. A few weeks elapse and the wife, overwhelmed by the domestic work, goes on strike. A neighbour comes by and advises her to change strategy, to work hard and by example coerce her husband into joining her. At the end of the story, the man ends up converted. He succeeds in building a perfect homestead which has all the trimmings exemplified in the display booths at the back of the performance arena.

From this, one may conclude that the main difference between the *Balijja* domestic chores text and that the text articulated in the women's play was the addition of a money-generating component to domestic work. Women's ability to generate and control money was perceived as a transformatory tool expected to effect women's independence from men. The final stanza of the song by the Vvumba women in fact threatened husbands who prevented their wives from engaging in money-generating activities. It stated in part:

If you dare say
"Women will not work" (for money)
We shall divorce you . . .
And leave you gazing

In the final scene of the above-mentioned women's play, a subtle gesture marred the otherwise picture-perfect ending. The husband had just returned from shopping and gave his wife a *busuti* (traditional outfit that is made out of a seven yard piece of fabric and is complicated to wear) while he started to put on the *kanzu* (a simple tunic) he had purchased for himself. He tried, but failed to locate the sleeves to his tunic and had to be rescued by the wife who had successfully completed putting on her *busuti*. The woman was clearly in control looming over the husband who was made to appear clumsy and incompetent despite the fact that he was governing the household as was the traditional custom!

Although some songs commended co-operation between men and women, the majority of them tended in essence to dethrone man from his traditional seat in the household. This phenomenon was reflected in the songs such as the one by Busubizi Club that focused on prescribing the structuring of household objects in the homestead. This song, already referred to above, while addressing itself to a female audience, included in

Socialization into Gender Specific Roles

The above song, for instance, invoked and undercut the text of matrimony in the Kiganda culture through the traditional song game, *Balijja*, one of the most commonly performed texts of matrimony in Kiganda culture.[3] In that game, a newly-married woman who silently rebels against servitude in marriage is discarded by the husband.

Balijja is actually a children's song game which is designed to teach young children institutionalized gendered responsibility within marriage. The game is usually performed in a circle. The children crown a king and a queen then sit in a circle. One of the children, acting as a bachelor, skips around the inside of the circle complaining about his destitution as a man without a wife to perform household chores for him. With the other children responding with the chorus "*Balijja*," the selected bachelor walks in the inside of the circle and with a self-pitying stance chants about how he trudges along like a poor little lamb of the palace having to cook for himself and to carry out all the other menial chores reserved for wives. When he arrives in front of the king he kneels down and asks the king to give him a wife. The queen looks silently on as the king gives a wife to the bachelor.

The bachelor continues chanting and walking around the circle followed by his newly-wed wife who does not utter a word. By the time the husband goes round the circle and once again arrives in front of the king, the marriage has gone sour. He complains bitterly that the wife he was given refuses to cook for him and that when she does cook, the food is either half cooked or burnt. The first wife is dropped out of the circle while the man begs the king for another wife. The king points to a new woman who silently starts on the matrimonial journey silently following the husband. The game goes on till the king runs out of women to give away. The moral of the story here is — "women, serve your husband or else you will get divorced." At a deeper level, (generally not understood by the children performers) while the wives do not utter a word during this game, they utilize silent protest to successfully subvert the ascribed roles by refusing to perform the prescribed domestic chores.

A play by women from the Lake Wamala area further re-enacted and reformulated the *Balijja* text. Most importantly, this play moved the text from the children game context to an adult marriage context in the centre of the village. When the play opens, the newly-married woman is sitting in the centre of a dirty compound in front of a poorly-thatched house surrounded by high grass (the type of bachelor dwelling described in the women's songs). The man is seen spelling out to the wife the rules that a man, and the only being on earth. Kintu is a primitive being whose diet consists of only cow dung and cow urine. Nnambi ascends back to heaven with Kintu to ask her father's permission for her to marry Kintu. After passing a series of tests ordered by Ggulu, Ggulu agrees to the marriage. Ggulu orders Nnambi and Kintu to leave for earth immediately while *Walumbe* (death), Nnambi's evil brother is absent from heaven.

On the way down to earth, Nnambi discovers that she has forgotten to take millet for her chicken with her. She disobeys her father's orders not to return to heaven and coerces her husband to return with her to heaven to retrieve the millet. *Walumbe* sees Nnambi and Kintu and insists on descending to earth with them. On earth, Kintu and Nnambi beget children. *Walumbe* soon starts killing these children. Efforts to expel Walumbe from earth fail. Kintu resigns himself to this fate saying "if *Walumbe* is determined to continue killing my children, let him do so. He will never exterminate them because I will always beget more." Thus, up to this day, the Baganda actually address themselves as "*Bana ba Kintu*" (children of Kintu). Nnambi, on the other hand, through the act of disobedience is remembered as one who brought death into the world. Benjamin Ray's (1991) extensive research on this myth documents how male interpreters of the story of Kintu ascribe authority of the primordial household to Kintu and state that women occupy an inferior position in Kiganda society solely because Nnambi is believed to have brought death into the world (Ray 1991). In addition, Nnambi's return for the chicken's millet is used to justify the Kiganda tradition of prohibiting women from eating chicken and eggs. In the festival song under discussion the women articulate the perceived liberation of contemporary Baganda women utilizing the fact that women are now able to eat chicken and egg. The Baganda's reliance on the Kintu narrative as a charter for gender norms presents opportunities for intertextual interpretations of the Mityana women's songs and performances.

When the women, singing about their demonstrated skills and knowledge of maintaining their households, name themselves "founders of the nation," "mothers of the nation," "light of the world," it is very possible that they are at least subconsciously dialogizing gender ideas embodied in the Kiganda founding myth. They are in fact creating counter-hegemony discourse in response to the culture's anti-feminist founding myth.

Furthermore, I believe that rural women's successful participation in the recent guerrilla war that liberated Uganda from the "Obote II" dictatorship has been a powerful catalyst in the women's recent aggressive search for gender equality. Their role in that war has been praised in radio its prescriptions architectural instructions for the construction of the ideal house. The song gave specific description for the type of roofing, windows and other parts of a building. This act, in a rural context, could be construed as subverting traditional male authority that is vested in the husband's responsibility to provide a residence for his wife.

Furthermore, the structure of a number of the songs characteristically contained three distinctive parts in varying sequences. One part described aspects of the objects in the exhibits; another explicitly or implicitly established female identity or authority in relationship to the making and ordering of those objects; while the third part depicted unemployment or joblessness among men and its effect in reducing their domestic power and influence. It can be concluded from the above that the songs of the club women contained tactics for recreating female images in the context of their relationship to the objects they created in the household.

Abakyala Musitule Amaaso is a song that provoked angry responses from the men in the audience. It started by narrating the oppressed history of the women's foremothers and compared that history to what the women considered their own more liberated status. The song recalled, for instance, how "women of long ago" in contrast to today's women were prevented from eating chicken and eggs and were thus deprived of access to good nutrition. The last part of that song became a biting attack on what the women termed men's liberty to be indolent, greedy and irresponsible while leaving both food and money-generating activities to the women. In addition, the song depicted the man as a drunkard who eventually drifts into nothingness. Thus, in this particular song structure, the text totally dethrones man from a position of authority in the household because of his lack of industry. The women's own industry, particularly as evidenced by the skillfully made objects on display in the booths, embodied *achieved* power and provided the basis for the women's texts in the songs to claim or demand authority in the household.

By attacking men's authority and power in the club songs, the women were, by implication, questioning and eventually rejecting the source of men's traditional power and authority over them. This source of power, according to Kiganda tradition is traceable to the Kiganda founding myth, the story of Kintu through which the authors or interpreters of the myth ascribed power to maleness and subverted the power of the primordial mother, Nnambi. Such constructions of power in gender relations have been addressed by multiple scholars including Henrietta Moore (1988).

This story of the origins of the Baganda starts when Nnambi, daughter of *Ggulu* (the sky God) descends to earth to take a walk and meets Kintu, songs and in other public media by both men and women. Women at the grassroots are now able to participate in local elections for positions in village governance. This political and social environment has sensitized the male population to the need for more significant roles for women in public affairs. This has substantially contributed to their ability to conduct the level of gender discourses as those represented in the July 1992 Mityana Women Club's festival. Women's empowered public voices are now penetrating both their cultural and material productions. Investigations into issues of contemporary gender politics such as the one by Mbulelo Mzamane (1992) is crucial to projecting the problems facing the African woman's involvement in national politics. As stated in the introduction to Toril Moi's (1985:4) *Sexual/Textual Politics*, "effective feminist writing [is] work that offers a powerful expression of personal experience in a social framework." As demonstrated in this discussion, the Mityana women fit this conception of feminist discourse. They perform their lived experiences within a space authorized (by men) while simultaneously creating (unauthorized) feminist discourses that criticize patriarchy and aim at producing actual changes in the gender power structures.

NOTES

1. This is an extract from a forthcoming book by the author titled "Narrating gender: orality, female space and the Kiganda Radio Song." It consists of articulations of initial findings based on my research into songs and performances produced by the women of Mityana women's clubs from around the districts of Kibuga, Mubende and Mpigi, Uganda.

2. See also Kofi Agovi's research on women's songs in "Women's discourse on Social Change in Nzema (Ghanaian) Maiden Songs." Here, Agovi offers a thought-provoking examination of women's songs on social change as an expression of feminist consciousness within an indigenous female group performance tradition.

3. See extensive discussion of the feminist text in this game in my article: "Intersections: Gender, Orality, Text, and Female Space in Contemporary Kiganda Radio Songs."

PART IV

PROSPECTS

Chapter 13

COMBINING MODERN AND TRADITIONAL MEDIA FORMS IN PROMOTING DEVELOPMENT OBJECTIVES IN AFRICA

Njoku E. Awa

The purpose of this chapter is threefold. First, it explores the pessimism with which indigenous communication systems in Africa are viewed. It answers the question: What accounts for the denigration of traditional African modes of communication? Second, it shows how, even in a media environment saturated with Western information technologies, traditional media still have a role in contemporary Africa. Third, it demonstrates the importance of combining the African and Western media systems in producing and delivering development communication. The two systems can be complementary and can accomplish much more than either system on its own. The chapter thus proposes that programs for development change in Africa should explore the most effective ways of combining them.

The Roots of Pessimism

Western media were first introduced to Africa around 1800 with the publication of the *Royal Gazette* and *Sierra Leone Advertiser* in Sierra Leone in 1801 and the *Cape Town Gazette* and *African Advertiser* in South Africa in 1800. These newspapers, published in European languages for expatriate readers, served as official mouthpieces for colonial governments and provided trading information for commercial interests. The first newspaper of note using the local language was the handwritten *Iwe Irohin*, published in the Yoruba language in Nigeria around 1859 by missionaries seeking to convert Nigerians in greater numbers.

Early colonizers and missionaries did not, by and large, understand the native cultures of Africa or appreciate the complexity and value of traditional modes of communication. To help rationalize economic and cultural exploitation, Europeans had to perceive themselves as a superior race and culture and, conversely, to view Africans as primitive and inferior people. English missionaries set up schools modeled on Indian prototypes which, as Lord Macauley wrote, were designed expressly to create "a class of persons Indian in blood and colour, but English in tastes, in opinions, in morals and in intellect" (quoted in Ashby 1964: 2).

This sort of blatant racism eventually fell out of fashion, but the primarily on traditional and indigenous forms of communication. These include verbal, action, aural, and visual manifestations such as drama, histories, proverbs, public rituals and legends, accompanied by flutes, gongs, and drums — and they are often participatory in nature (Awa 1989: 313). By participatory I mean that they may require interaction with an audience. Indigenous communication thrived before Western intervention and today has reached a high degree of development, as Okam (1978: 242) states:

> Long before the Roman, Arabic and Amharic letters and scripts were used in Africa, proto-literary creativity flourished on the continent in the form of tales, legends, myths, epic narratives, proverbs, invocations, limericks, recitations, ballads, chants, songs and dramatized religious liturgy. This tradition was oral, transmitted from one generation to another by word of mouth.

When people interact face to face, the communication usually flows in two directions. A dialogue entails feedback, participation, and involvement of more than one party. In other words, it is an exchange — in stark contrast to the electronic media, which broadcast their messages in one direction to a passive audience stripped of the ability to respond. Even newspapers, with the exception of the editorial page, cannot carry on a discourse with their readers. As was true when Europeans first introduced mass media to Africa, content, perspectives, and the delivery of messages are controlled from a central source. Such a system slights the participatory aspect of traditional African forms of communication. Hachten and others appear myopic in overlooking this cultural clash. Their oversight would be harmless if, indeed, communication was not important for development aimed at improving people's quality of life. But communication, whether through mass media or traditional channels, has a role to play in changing people's lives. How can communication play its role *effectively* to benefit the poor and disenfranchised.

A Role for Indigenous Media

African oral tradition has a legacy that extends beyond the coastline of the continent to far-off shores and even through the airwaves of industrialized countries. Popular stories in American culture have come directly from old African legends and tales. For example, Walt Disney's *Song of the South* employs a cast of characters led by Brer Rabbit, a cultural descendant of Ananse, a rabbit character from African legends known for his trickery and quick wit. The story evolved to its present form being passed down perceived superiority of Western society persisted subliminally in the paradigm of development that emerged during the 1960s. Indigenous media were overlooked as "legitimate" channels for change, whereas Western mass media were viewed as essential for development.

Daniel Lerner (1958: 56), one of the foremost proponents of this position, maintained that what had previously been referred to as the Western model of development was indeed a global model: "Once the modernizing process is started, chicken and egg in fact 'cause' each other to develop." For Lerner (1958: 46), urbanization, literacy, and widespread use of mass media formed a pivot around which modernization revolves:

> Everywhere, for example, increasing urbanization has tended to raise literacy; rising literacy has tended to increase media exposure; increasing media exposure has 'gone with' wider economic participation (per capita income) and political participation (voting).

These three engines of change in the modern world, he postulated, are fueled by physical or geographic mobility (traveling or settling outside one's village) and psychic mobility (vicarious experience via media exposure). Lerner implied that as countries modernized, their mass media would concurrently develop along the lines of Western mass media — an assumption that precluded the use of indigenous oral systems that had worked effectively for centuries (See Fig. 13.1).

Sector	Media systems	Oral systems
Socioeconomic	Urban	Rural
cultural	literate	illiterate
political	electoral	designative

Fig. 13.1: Urbanization, Literacy, Voting, and Media Participation Matrix.

The failure of mass media to take hold in Africa has been cited by many as a symptom of the poor showing of development efforts on the continent. William Hachten (1971: 51, 52) cites five factors contributing to this failure: dictatorship or one-party government, socioeconomic constraints, lack of capital, no protections for freedom of the press, and corruption. Many scholars find some merit in Hachten's analysis, but many also feel he overlooks the factor of congruity in matching the medium and message to the audience.

Though mass media are not new to Africa, most Africans still rely

Now, from Nigeria to Swaziland, there are those in Africa turning to theater to teach basic health care, family planning, conservation and problem solving. It has been particularly effective in the countryside, where there is no electricity to run movie projectors and television (*New York Times*, April 4, 1983).

The reasons for the popularity and effectiveness of this mode of communication for development-oriented messages are self-evident. First, it is transactional and highly participatory, allowing participants to work with new information through a familiar and comfortable format that also allows communicators to modify their messages in response to verbal and nonverbal audience feedback. For example, Bame (1990: 134) reflecting on his study in Chapter 4 writes about the use of traditional drama to communicate family planning in Ghana:

> We chose the Concert Party Plays because they not only effectively combine visual and oral effects in driving home their message but they are familiar to, and highly popular among Ghanaians of all walks of life but especially among the rural illiterate folks who were our target population.

Second, the oral system closely matches the structure and worldviews of the African societies that use it. Again in the case of Ghana, Pickering (1957: 178) states that "village drama is the most truly Ghanaian audio-visual aid, depending as it does on a nation-wide aptitude and liking for drama and by its intimate relation to local custom and tradition." Folk media are therefore capable of giving local relevance to communication by adapting it to local sociocultural systems. This adaptability may add an important degree of credibility to messages.

Consideration of media channels also brings up questions about cultural imperialism. Even if an African country were fully to adopt Western-style mass media with an attendant audience, what would be the cultural costs? Daniel Lerner's assumptions on development pay little attention and give little value to the cultural dimensions of a modern communication system. Development is not simply a question of innovative technologies and economic development. It must also take into consideration cultural and environmental factors if the end goal is to improve the quality of life for people. As Luke Mhlaba (1991: 210) writes in reference to Zimbabwe, the obstacles to the emergence of a mass-based development culture

> lie in the cultural alienation of the people, which consists in the imposition of arbitrarily defined institutions of socio-political organization, values and, even more importantly, ineffective means of communication between the masses and the elite.

through the generations by enslaved African-Americans. Its survival clearly demonstrates the strength of the oral tradition, even under the most adverse conditions.

Another example of the oral tradition is the "call and response" found in the gospel music of African-American churches. This form stems directly from traditional African drama and the role of the raconteur. This same gospel music evolved into "the blues," which provided the musical foundation for the rock and roll music that dominates radio air time throughout the world and supplies the popular and influential MTV with its *raison d'etre*.

The apparent influence and resilience of African oral tradition is impressive, but of greater import to the new generation of development practitioners is its theatrical and dramatic forms which encourage audience participation rather than restrict interactions to the key players.

In the past, ideas were spread from one village to another by a skillful raconteur, or practitioner of oral literature. The breadth of common experience was such that the audience would frequently participate in the refrain of what was ordinarily a communal drama (Awa 1980). This approach combined learning and entertainment, an approach now used increasingly by Western educators in developing their curricula and teaching materials. Such a participatory approach has the potential to enlarge the sensibilities of all those involved. It makes participants active in their own education by encouraging response and reflection. This kind of communication has been recognized by various scholars, including Paulo Freire, as a key to empowerment, which allows local people to break the bonds of silence and ignorance (Dejenne 1980: 4). This is a most important step on the path to self-reliance and development.

Starting in Ghana and Gabon, development communicators went back to their communication heritage to pursue "educational goals pertaining to health, literacy, or the control of cocoa tree pests, by means of a simple, rather spontaneous form called 'vernacular' or 'village' drama" (Doob 1961: 78). More recently, development programs in Botswana, Nigeria, Ghana, Zimbabwe and Zaire have sought to revive villagers' interest in drama as an instrument of educational development. Kidd and Bryam (1977: 1) have described a theatre festival in Botswana "to promote participation and self-reliance in development by bringing people together to discuss their problems (reflected in performance), agree on changes that need to be made and take action." An Associated Press report on the popularity of this medium notes:

directives issued from the top, from one all-powerful leader. In response to these sessions, programs often address issues brought up by farmers in their local forums; listeners take a meaningful part in their own development.

The forum approach also keeps the broadcasters and planners on track. Without this crucial feedback, the communication tends to become "top-down," relying on the instincts and best intentions of the planners rather than assimilating the real-life conditions and experiences of farmers at the grassroots level. Feedback enables programmers to address issues that are relevant to their audience.

Although the Farm Radio Forum was first developed for use in a high-technology nation, it has been successfully adapted for rural adult education in India. On the basis of that success, radio forums were introduced in Africa in the early 1960s. Although training programs were conducted for countries using either French or English as the colonial language, the anglophone nations, specifically Nigeria, Ghana, Malawi, and Kenya, made the most effective use of the radio forum model.

Tanzania has used the radio forum model in a slightly different way, but with considerable success. Rather than establish an ongoing program that over time will cover a wide variety of topics, Tanzania has emphasized a campaign approach during which radio broadcasts are coordinated with local study/discussion groups that focus intensively on a narrow topic for a limited and relatively short period of time. Tanzania's 1973 "Man Is Health" campaign reached, it is estimated, nearly two million people with radio messages that promoted the construction of pit latrines and the clearing of vegetation from the immediate surroundings of houses. Those participating in the campaign were judged to have increased their health knowledge twenty percent over those in control groups (Hedebro 1979: 96).

Recently, a different type of Western influence came to bear on a unique development campaign in South Africa. An indigenous music group, The Prophets of the City, began a series of Rapping for Democracy concerts to help promote political participation among the country's African majority. It adapts a musical form born on the streets of America's Afro-American communities to a traditional form — the call and response — to carry a new message, the importance of democracy and the need to vote.

The reason why we gotta vote,
the government left us broke
in each and every way.
And now we have to have a say.
Let us go for a better day.

Thus, the use of folk media is necessary to achieve some kind of cultural stability in an industrialized context. Change that threatens cultural stability can occur only gradually, as people agree to alter their value systems.

The main limitation to using indigenous communication forms more extensively is that they are very labor-intensive. For example, indigenous drama can reach only one village at a time, and the number of annual performances may depend as much on the endurance of the development team as on their skills. Although it is an effective medium, it may reach too small an audience to effect change on the massive scale needed to set the countries of Africa onto a fruitful development course.

Traditional Communication and Mass Media: Finding the Right Mix

Can traditional African communication and Western mass media be adapted to each other and harnessed in the service of development objectives? Although efforts in this direction have been relatively few and far between, several notable cases provide encouragement. Some future approaches, if carefully adapted to the African context, may also hold great potential.

Perhaps the most encouraging example of modern media working together effectively with traditional modes of communication is the radio forum. The Farm Radio Forum model originated in Canada in an effort to facilitate two-way communication between development planners and rural people after the drought and depression of the 1930s. The forum model relies equally on scheduled radio broadcasts and on communication to introduce and encourage rural populations to adopt new or appropriate technologies. Extension officers organize groups who come together to listen to a scheduled broadcast on a set topic. Discussion plans and materials are distributed to the officers in advance of the broadcast to facilitate interaction among the participants. At the conclusion of the session, comments and reactions from the group are forwarded to the broadcast producers for follow-up.

This approach incorporates the use of radio, the most accessible medium to rural populations in Africa, and it relies heavily on discussion among the participants to produce interest and results. Feedback from the villagers to the broadcasters empowers the participants by giving them a voice that is listened to. The radio forums are culturally compatible with African ways of making decisions, being similar in structure to the collective modes commonly found in African communities. Communication flows "horizontally," emphasizing group discussion and decision-making over counterparts. Second, these were branches of an oppressive military government propaganda and law and order enforcement machine which was required to *disseminate* information and not to share or discuss it. Third, they were quasi-intelligence organizations for whom secrecy was paramount. Fourth, not many people would want to be associated with these political groups. In fact, members of the groups were often treated as *paraiah*. The people's desire to distance themselves from these groups was displayed by an elderly woman who pointed at them as the "government's people." Actually, the fear generated by these truly Athusserian organs of state was so intense that it generated a "culture of silence" within the nation.

As already noted, at official assemblies, the Chief communicates through the *Okyeame*, the spokesperson; but speaks directly to everyone in unofficial interaction. Ansu-Kyeremeh (1989) recounts how during his fieldwork he once bumped into a Chief in conversation with friends the former was visiting. Spouses spoke to each other freely, but each needed to choose words carefully when communicating with in-laws. An *Okomfo* (priest of an indigenous religion) and Elders were greeted (especially by women), with a bow or a genuflection; and men would slip their cloth off the shoulders in deference. Thus, excepting the Chief-subjects official communication, no rigid communication barriers, arising out of the above social differentiations, were observed. Spouses, parents and offspring, people of different religious faiths, males and females, and adults and children shared free, two-way communication with one another.[5]

Intra-Village Communication Patterns

That communication is the essence of village community was borne out by the intricate but clearly defined network of intra-village communication patterns that Ansu-Kyeremeh (1989) observed among the villagers he studied. Nine different types of indigenous and more than four Western technology-based communication systems were identified by him. The indigenous ones could be distinguished from each other on the basis of source of message likely to be channelled through that specific medium, the nature of the message, the target group for the message, the basic purpose of communication served by that medium, and the process and direction of the particular communication. On this basis four basic classifications could be made. They are: venue-oriented communication, events as modes of communication, the communicative nature of games, and performance-oriented games. Ansu-Kyeremeh (1989) found high participation and access rates among villagers in these indigenous communication systems which are described next.

Venue-Oriented Communication

One pattern of village communication was manifested in venue-based communication transactions. In this mode, certain venues in the villages served as meeting points where conversations naturally developed. At these gatherings, information was passed through structured or unstructured conversation among people present.

Venues where this pattern of communication was generated were identified. They included the Chief's or the Queen Mother's court and verandahs. Others were artisans' (blacksmith, carpenter, potter) workshops, *nkwankwaannuase*, *gyeduase* (tree shade), *kookoo aboe* (cocoa pod splitting venue), *nsadwase* (drinking place), *ahoroe* (riverside washing sites), and *afukwanso* (trails leading to farms).[6] *Nkwankwaannuase* is so called because it was a popular gathering spot for the youth (*nkwankwaa*) who together constitute a formidable political force and have their own chief, the *Nkwankwaahene*.[7] The *Nkwankwaahene* will always be consulted by the *Ohene* [Chief] in any thing which will involve the young men . . . (Obeng 1988: 30).

Meetings at these spots were informal and in small groups with the presence of people purely on a voluntary basis. Ansu-Kyeremeh (1989) cites examples of these gatherings on the senior *ahenkwaa*'s verandah; at a carpenter's shop; in front of the house of a village Elder where people had gathered to play the *oware* game; young women's hairdressing; young boys' meeting at the house of a teacher; and at blacksmiths' shops in various villages. A small village market was another type of venue. In addition Ansu-Kyeremeh (1988) believes that even the confines of a moving vehicle plying between villages served a communication purpose.

At these venues, people were involved in interactive conversation. Conversations centred around practical and useful information or focused on the importance of early vaccination of children. Many learned of goings-on in the villages at these venues. The meetings could be in large groups (sometimes involving the entire village adult population, especially when there was a formal summons or invitation).

Village assemblies or rallies may generally be summoned on behalf of a government official or a stranger by the Chief's authority (exercised in council) at any of these venues where convenient. In the past the information officer, the community development officer, and health education teams took advantage of this setup.

animators develop posing materials or codes (pictures, posters, slides, songs, etc.) depicting a concrete experience, relating in one way or another to a problem. The code, through its resemblance to real life, should be designed to provoke discussion from members of the community to discuss a particular problem. Group discussion leads to reflection on the problem at hand and strategies to address the issue. The code may be tried out by animators ahead of time with a trial group, to see if it is an effective catalyst for discussion and reflection. While traditional forms of communication may be used, Hope and Timmel (1984: 137) point out that "Certain films can serve as very useful problem-posing codes." Only through experimentation and careful consideration can one be sure. Yet this approach seems to be a viable means of combining modern mass media with more participatory modes of communication. It suggests the possibility of combining mass communication channels such as film, radio, and even television programs with extension work.

Social marketing may appear an attractive new approach to reaching the masses. It should be undertaken gradually and with great care in Africa if we are to avoid the failures of the past and build effective new communication channels in the future. To be successful, it will require the close cooperation and sensitivity of indigenous communicators and mass communication specialists to bring their talents to bear on creating strategies and messages that are palatable and persuasive to local audiences. It will also require the development and promotion of channels for feedback to communicators and participation for the listeners, readers, and viewers; and a perspective that places the needs and communication styles of the indigenous audience over the novelty of the technologies and the "flash and dash" of the programs and campaigns.

Summary

Thus, the pessimism with which indigenous communication systems are viewed in the African context does not seem to be grounded in any theoretical perspective that sufficiently justifies the exclusion of those systems as a means of achieving effective communication. And the fact that in Africa development communication objectives are often not met because of factors that are associated with excessive reliance on the modern media suggests a need to rethink the role of the indigenous systems. In considering the inclusion of the indigenous systems in future communication planning, one needs to consider not just their attributes but also the contextual socio-cultural chracteristics with which the systems are intertwined.

Chapter 14

TOWARDS EFFECTIVE COMMUNICATION POLICIES AND STRATEGIES FOR AFRICA*

Des Wilson

Introduction

Many African scholars, in recent times, have spoken of the need to have a truly democratic and virile communication policy for Africa. Prominent among these scholars have been Ugboajah, Nwuneli and Blake. Although they have not articulated their positions in exactly the same way, they all seem to advocate new communication polices and options which they believe should equally reflect a new order. Africa, they all seem to say, cannot afford to remain helpless and exploited by those nations which now possess the competitive advantages of modern technology.

It seems quite obvious that Africa has remained helpless in the communication sector because of certain factors. Her leaders often equate a longing for development as being synonymous with the acquisition of technology. This, they believe, can be achieved through what they perceive as technology transfer, a euphemism for a subtle acquisition of knowledge without trying. To the surprise of many of these leaders their expectations have not been met. It has not occurred to many that such expectations are unrealistic. There is, therefore, a crying need for a re-examination of Africa's indigenous systems which have not been fully understood nor studied. Such an exercise would lead to the isolation of those aspects of traditional media and the new media which could then be integrated for the purpose of establishing a more effective system of communication for the continent. This approach would lead us to a diachronic-synchronic view of the appropriate strategies to use for communication in Africa.

A Diachronic-Synchronic View of Communication

This chapter proposes that a historical perspective of Africa's communi-

*This chapter was first published in the *Africa Media Review* (1989 Vol.3, No.2, 1989, pp. 26–39) under the title "Towards a Diachronic-Synchronic View of Further Communication Policies in Africa." This review version is published under a new title "Towards Effective Future Communication Policies and Strategies for Africa." Acknowledgment is therefore given to the African Council on Communication Education, the publishers of *Africa Media Review*.

the present policies must be reversed to pave the way for a truly democratic and legitimate information system hinged on a trado-modern communication policy.

Communication as Culture

Communication is a word which is eclectic in its forms and is varied in meaning. It has therefore, been visualized as anything that contributes a meaning of its own, or any activity which is carried out by man within his cultural millieu. Fiske (1982:1) sees it as a multifaceted area of study which includes interpersonal talk, television, spreading information, the hair style, literary criticism, and many others. It is from such a broad spectrum of views that communication as a term has acquired its present image of being a melange of sociological and language-related activities. It is this that probably led Watzlawick *et al.* (1980) to say that one cannot not communicate. Whether in sleep or consciousness, man is known to communicate in so many ways. It is inconceivable to imagine a situation of sensory deprivation resulting from total blackout of living persons from events around them. Human being communicates as long as there is breath left in them. Thus, every culture has its varying forms and symbolisms of communication. Communication is culture it is a manifestation of the cultural norms of society. Through the language of presentation to the very symbols, themes, channels and media used, the culture of a specific group is reinforced; a new awareness is created and, sometimes modifications are made to existing norms. This monopolistic pattern of cultural generation in more technologically advanced cultures has led to cultural or media imperialism in the Third World. Technologically disadvantaged nations have in consequence capitulated economically and politically as a result of the seemingly superior communication and culture which is backed by powerful Western technology.

So as communication reinforces the cultural norms, creates a new awareness, or modifies existing patterns of behaviour, it does so through a certain role-set created for it by the very culture it reflects. First, it educates individuals within the culture. Thus, it operates as a regeneration system forged to maintain the commonly-shared norms of the people. Second, it provides entertainment and pleasure thus acting as a balm to the society. Third, it informs the society about events through the watchful eyes of its agents. It thus trains its eyes on every aspect of society from the king to the thing. Fourth, in Western capitalist society, it exists and operates to make profit in order to be able to sustain the system. This role distinguishes the cation system should be adopted to go hand-in-hand with those approaches which have been used over time. This approach will enable Africa to tackle her communication problems more realistically. This view sees communication as a cultural transaction and transmission which takes place over time. The technology for transmission of messages undergoes changes and is shaped and sharpened through the ages. Old processes are synchronized with modern technology as new ones replace some of the functions of the old ones. This process of synchronization of the old with the new shows the past acting as a guide to the future. With this, there is no age or cultural gap created by jumping from one form to the other without being part of the socialization process.

 .   Thus a diachronic-synchronic view of communication places in focus a traditional and modern communication model which Wilson (1987) suggests for developing societies. This view encourages the study, classification, and examination of traditional communication and practices as well as their changes over time. The traditional become the roots of modern communication while the present ones act as appendages or outgrowths of the past. This calls for a closure of the cultural gap between traditional and modern societies. In practical terms, this means the marrying of traditional communication practices with modern practices in appropriate proportions, an establishment of an interface between the two in an equal linkage without one suffering loss of status. This approach which should take into account the rate of growth of the society should be gradualist.

      The traditional communication processess have been ignored for too long for what often appears to be more efficient but ineffective, faster but limited, heterogeneous but alienating, time-saving but expensive modern processes which are after all, culturally threatening to the process of democratization and legitimization. The new process should aim at returning information dissemination process to the people because these processess belong to the people.

      Communication as an industry has now become an elite enterprise richly touted in expensive boxes (television and radio), and in costly news sheets (newspaper and magazine). The processes of democratization have been reversed for most societies since those for whom information is meant never gain access to it. Even when they do, they treat it with distrust because the political organization which controls the system is alien to them and the leadership is often seen as an offshoot of the dehumanizing colonial experience. Political leadership is often established through brazen fraud and thus forfeits the legitimacy which it demands but which it lacks. It is, therefore, inevitable that in order to return Africa to its period of innocence

Furthermore, the arts and music of the people are often expressed through the traditional channels in society. For example, sculptures and art pieces speak as eloquently of the culture of the people just as their architectural designs say a lot about their tastes and habits. Moreover, the scientific and technological developments of any nation are communicated through communication channels which have their roots in the culture of the people. Scientific and technological devices have made modern communication possible through various devices which, in turn, have become part of the social norms of the people.

Then one may ask, what is communication if not a young African saying "Good morning" to an elder with an appropriate gesture of respect? Communication and culture lie in the individual's use of proxemics, chronemics, kinesics, haptics, occulesics, objectics, numbers, colours, olfactory and gustatory senses, symbolic display, semiotics, picture, dressing and other forms of symbolism to express relationships. Some examples should suffice here.

The distance between two persons in an open place could be used as an indicator of the relationship between them. A student's early or late arrival for lectures or other academic activities may be an indication of a personal attitude, perhaps deeprooted in his cultural environment. The way a student walks into the class, perhaps wriggling her waist and rolling her eyes with a knowing smile on her face, speaks more eloquently than verbal expression of those actions.

Girls and boys are often conscious of the way they look — jerry curls, the mini, midi, or maxi-skirt, jeans, blouse cut deep-down in front, skirts (slit or plain), Lady Di make-ups, all these underscore the power of expression among the youth. If all these are taken to be personal mannerisms, then because of the frequency of their occurrence, one may say that they are the personal mannerisms of the group or culture.

Colours also play some part in cultural communication. They are relatively speaking, free from linguistic barriers and have the advantage of the speed of impact which many other forms lack. Although there are colour universals, some cultures differ in the application of colours. For example, some colours have been given racial overtones, especially the colours black and white. It is a fact that Africans of brown skin never called themselves black until the "white" man started calling them so. All the negative connotations given to black seem to be all racially motivated and, since the English language found a convenient colour label for the white race, it became necessary as a fact of colonialism and imperialism that the thinking press system of the socialist countries from those of the capitalist ones as well as the press of the Third World from the rest of them. With these we are confronted with the inevitability of viewing communication as culture. It is reasonable to agree with Fiske (1982: 2) that "the study of communication involves the study of the culture with which it is integrated."

The Cultural Content of Communication

Communication theory has broadly borrowed from the fields of anthropology, psychology, history, sociology, linguistics, music, religion, mythology and the physical sciences. Being derived from such a motley of disciplines invests the communication field with the identity of a culturally derived field, that is, its roots are firmly set or fixed in the culture of the people. It is also seen as a multi-disciplinary field which strongly reflects the music, arts (sculptures), architecture, religion, ideology, politics, social norms, science and technology of the society. It is a field which has status because it derives its strength from a variety of sources. The communication industry which sustains the business of communication is as varied and farflung as its different antecedents. The industry comprises the electronic gadgetry, wood and paper, machinery, chemicals, educational institutions, computers and a whole complex of trades and professions. It is no wonder then that the cultural content of communication is like a mirror of human activities.

The over-riding communication message reflects the ideology and politics of the dominant powers. Through communication the long dead events of history become the beacon stones and searchlights of today and the future. Therefore, an understanding of the past leads to a greater awareness and better grasp of the strategies of modern living.

In our time, religion has become the big cultural or communication activity in many parts of the world. Televangelism is a significant feature in many state-run broadcast stations. A few years back, the conflict created by different religious interests led to a government ban on radio and television broadcasts of religious programmes outside the regular days of worship of the two dominant religious groups in Nigeria. Religious activities, in spite of these difficulties, are still big media events and almost a permanent fixture in the Nigerian cultural spectrum. The horizon and landscape are peopled with as many religious groups as there are countries in the world. Religious programmes on television and radio in many Nigerian states, apart from reflecting traditional practices, are also channels for expounding the cultures of the originating countries of the particular religion.

The source is the village council of elders and chiefs, at the head of which is the village head. The gongman (town-crier) is part of the gatekeeping as a broadcaster or reporter. He uses the gong, skin drum, metal or wooden drum, flutes and other instruments which he beats or blows to win or attract the attention of the potential audience member in their various homes. The sound produced by the instrument acts as a signal. He then broadcasts his message which may also contain all elements of modern radio broadcasting with the additional advantage that some of the audience members may see the "broadcaster" in person. Generally his audience is in close contact with him and readily provides feedback, individually seeking explanation to issues that may not be explicit at first. The gongman may provide instant answers to the audience feedback or may take the feedback message to the village head or the council where observations or objections to directives may be discussed or reviewed. This immediate talk-back system puts the traditional communication system ahead of its modern counterpart. There are also similarities between the traditional communication process and the modern system. First, in many African countries, the broadcasting system comes directly under a government department or ministry headed by an appointee of the president or prime minister. The minister reports to the executive council or whatever body is responsible for executive decisions. This council also appoints the chief executive of the broadcasting organization who in turn reports all the activities of government to the people. So does the traditional media system except for the fact that the modern media system lacks immediate feedback channels.

Modern Communication Technology

Modern communication systems are too expensive and sometimes too predatory to be ignored. There may still be several localities where television and, sometimes, radio sets are still unavailable today. But this is not on account of the inability of transmission facilities to reach them (as the case may be for a few), but because of the needs of the stomach exerting a stronger pull on the resources of the people than those of the eyes.

Modern communication facilities from satellite systems, data links, computer machines, printing machines, transmitters, cameras, compugraphic machines, word processors, picturephones, among others abound. They can be purchased especially by poor countries at sums of money quite beyond their national budgets. They are available to facilitate or make possible radio and television broadcasting, newspaper, magazine and book production, as well as supporting services like cable television, of the people equally had to be subverted through the tripartite god of gold, glory and God.[1]

Forms of culture may be modifications of existing patterns of innovations which bring about a new awareness; they may be mere forms of reinforcement. The cultural content of communication is therefore diverse and diffuse. Thus one cannot communicate what is not in existence either in the sciences, myths, religion or practice of the people. What is communicated is. What is not is not and cannot be reasonably communicated. Obviously, communication media are used as vehicles for transmitting culture. Therefore, culture is their message.

Traditional Communication Technology and Strategies

The technology which works the traditional communication system is simple and depends extensively on some sonic devices. Wilson (1988) describes these devices as instrumental media. They comprise instruments also associated with music, namely, idiophones, aerophones, membrano-phones and chordophones. These devices are used independently to transmit messages or are employed as part of certain communication modes or strategies to effect the exchange of information. These modes include: the demonstrative, ichnographic, institutional, visual and extramundane. The instrument used under these modes are beaten, blown, struck or plucked to produce sounds (signals) and messages whose clarity and efficacy may be dependent on the dexterity or expertise of the traditional newsperson, and on the nature of his message. The sound (signal) produced by the device acts as an attention-directing signal before the actual broadcast is made. For example, the wooden drum, as Alexandre (1972: 113) shows, when beaten by the newsperson, "reproduced . . . the rhythms and tones characteristic of the sentence he wants to transmit." Wilson (1988:9) also points out that

> most modes and forms display the capacity for multi-social functions and the communication message, the ability of the medium to get the message across to the audience in good time, and by cultural prescription of each society . . ."

The technology used in traditional communication is cheap and, at times, acquired at no cost to the community. The communication process is also relatively simple. the traditional communication system operates in basically similar ways as modern media systems in many African countries. The message originates from an organizational source which is part of the political system, inseparably linked by the mutuality of their roles in society.

they found them and then proceed to impose them on their countries. As pointed out earlier, communication is culture just as culture is communication. It is abundantly clear to many observers in the Western world as well as their agents in the Third World that the forms, structures and socio-cultural conditions under which the media operate in the West cannot be replicated in Africa for very obvious reasons, one of which is that our socio-cultural conditions are different. It is, therefore, in the interest of the dominating powers to continue to sell these instruments of power and culture to a people who, by viture of their disadvantaged positions, are unable to free themselves from the stranglehold of the more powerful culture.

Furthermore, socio-economic problems are also tied up with the above. Most African countries do not have the hard currency to pay for the equipment usually offered for sale by the dominating power. In order to overcome this problem, loans are arranged to further entrap these poor countries in the web of lending nations. The way out of this predicament is perhaps to limit the materialism of the leaders of these countries to those things which their countries can conveniently afford.

The other problem confronting many African countries today is that of political instability and diversity created by the 19th century European political chess game which gave Africa the variegated colour it is wearing today. Thus every dependent African economy is vulnerable to the political power play in Europe, America, China or Russia. Each power base tries to foist its own stooge on the people first by trying to create the society in its own image, and funding local Turks, especially restless soldiers, who find routine barrack experience monotonous and boring. Needless to say, every budding African soldier sees himself as a potential head of state. Herein lies the origin of incessant coups d'etat with the accompanying allegations of corruption and abuse of power against those unseated. Such incessant political changes bring with them frequent policy changes as well.

A very serious communication problem for most people in Africa is that of literacy and the use of foreign languages in the mass media. Ironically, foreign languages sometimes provide the only medium of reaching the majority of the literate public across cultures. But this is often taken to mean that such a minority public is greater in importance than the predominantly "illiterate" population.

But since there are numerous local languages in Africa and since the foreign languages (usually of European or Arab extraction e.g. English and Swahili) are often the only means of reaching the educated across the various linguistic barriers, it may suffice for now to continue to use them videotapes or videocassettes, phonograph records, and audiotape and cassette production.

These are all powerful instruments of communication for the technologically advanced world but they become liabilities in Third World countries. The hardware may be available for purchase but it is not easy to purchase the knowledge which may make its maintenance possible. The cost of buying the hardware (usually outdated by some five years or more at the time it is purchased) is always prohibitive. The maintenance cost is always even more outrageous. Thus the situation sometimes arises when the buyer country has to abandon the equipment to buy a new one because of the prohibitive cost of spare parts and maintenance. This is one area of drain in the resources of the Third World countries. The present situation calls for a new approach to communication policies in Africa in particular, and the Third World in general.

Problems of Communication in Africa

Africa is today saddled with multifarious problems from its perennial drought and food crises, political instability, economic crises, neocolonial and imperialistic subversions, and booby traps created by erstwhile colonial powers, to media and cultural imperialism now made easy through international satellite transmissions. These problems have arisen from a number of factors.

First, most African countries have very weak technological bases which cannot sustain them in the present century without external support especially if the philistine attitude of the leadership must hold sway. The false taste of the ruling class has contributed immensely to the political and technological woes of these countries because of the lack of supportive local "duplicating" interests in the acquired technology. Thus the superstructure, created largely on account of hardware fascination, is not an enduring one, hence the widespread decay of public property all over the continent.

Close on the heels of technological weakness is the issue of media and cultural imperialism. Many of today's African leaders have had some form of education or orientation in the technologically-advanced countries. Their exposure to the media in those societies sometimes leads them to propose similar systems for their countries. Even people with only a few days or weeks exposure to Western media systems are often the most ardent propagators of this form of media imperialism. They often try to take the media and their hardware out of the social and cultural contexts in which until literacy in the local languages has been increased. This seeming lack of an indigenous language to reach most of the people is responsible for some of the distortions encountered in the communication activities of African countries.

Finally, the present system of information flow between the rich and poor nations is the one-way flow. Since the dominating powers have the means of projecting themselves and their culture, the culture of the poorer nations is often distorted whenever it is portrayed. Exhibition of African culture is rare and, therefore, Western culture is frequently transmitted to poor nations through satellite systems belonging to and controlled by rich nations. This reinforces media and cultural imperialism and, hence, the poor countries are hardly heard or seen at the international level outside the United Nations and its agencies.

All these problems point to the fact that there is an urgent need to review the communication processes and reassess Africa's development needs and goals. Perhaps something close to a moratorium on all cultural, technological and media hardware considered to be of little value to the national objective should be declared for a period of at least twenty to thirty years. Then a renewal of faith in Africa's drive could begin to build on whatever her indigenous scientists and technologists can produce.

Summary

This chapter has viewed communication as the expression and reflection of the culture of any society except in those cases where media and cultural imperialism have taken over. It stresses the need to strengthen the existing communication systems through establishing an interface between the traditional and modern (Western) forms of communication. It posits the view that any policy which does not address itself to this dualist approach is bound to compound Africa's communication problems even more.

NOTES

1.   Professor Ali Mazrui's comment in the TV documentary "The Africans."

Chapter 15

INDIGENOUS COMMUNICATION IN THE AGE OF THE INTERNET: FROM THE VILLAGE CONVERSATIONAL SPOT TO THE INTERNET WEBSITE

Kwasi Ansu-Kyeremeh

In the years following the completion of the earlier chapters of this book in the early 1990s, Africa has moved closer to the second millennium and also into a globalized world of sophisticated information and communication technology (ICT). Communication technology has so pervaded the continent that there is not a single African country which has not been penetrated by the technoligcal media. Many African countries are now wired to the world of the information superhighway operated through the Internet system of communication technology convergence. Some of them have established homepages on the Internet. In Ghana, television stations (*GTV* and *TV3*) and radio stations (*JOY FM, Groove FM* and *Radio GAR*) are on the Internet live. Several newspapers including the *Daily Graphic, Free Press, Ghanaian Chronicle, Independent* and *Statesman* have websites.

Thus, indigenous communication systems (ICS) appear more anachromistic now than in the early 1990s. The first three chapters of this volume may be considered as belonging to the old order.

However, as is often reiterated, there are large sections of African populations which still depend on the ICS in meeting their information needs. In fact, there application of the ICS featured prominently in discussions involving the formulation of a national communications policy for Ghana.[1]

This chapter situates the ICS within the context of the globalization of communication technology. It attempts to establish similarities between the principles and dynamics of the ICS and those of the Internet. Among these are interactivity, individuality within the mass and asynchronicity. All of the new communication systems resulting from the fusion of the various technologies, have at least a certain degree of interactivity, something not unlike a two-person, interactive face-to-face conversation. The individual technologies are capable of reaching out to a large audience although their interactivity makes them more like interpersonal interaction. They, therefore, combine certain features of both mass media and interpersonal channels. Indeed, the notion of a visit to a website is akin to such as Stover (1984: 86) answer the question with the statement that communication need not be "technically sophisticated to be successful." In terms of intermediation, ICS are basically simple.

But simple or complex, communication machines and devices are invented to serve the purpose of people. These inventions are neither supposed to serve themselves nor are they to be served by humans — including their inventors. If machines are to serve people who exist within certain social, economic, political and cultural environment, then it can be expected that these aspects of human life would impinge upon the operation of such devicees.

Stover (1984: 82) actually believes it is important "to know what various communication means can do for a developing country, the advantages and disadvantages of each, and the problems associated with different types of information technology."

Perhaps, this is why long before Ghana's Fourth Republican Constitution sought cultural legitimation[5], a cultural policy enunciated in the mid-1970s had called for the integration of the indigenous forms of communication into emerging technologically sophisticated ones.[6] An adaptative mode of such integration is proposed by Awa. In Chapter 14, Wilson also advocates a diachronic-synchronic approach. Ansu-Kyeremeh (1997), though proposed an "indigenisation" approach to fusing indigenous communication and technological ones. The basic assumptive principle of indigenisation is that congruency of ICT with the local culture in which the indigenous systems are already embedded be the main criterion for considering a technological medium for integration.

At the same time as the 1992 Constitution was being drawn, an attempt at developing a "National Information and Communication Policy" for Ghana similarly recognized the importance of the ICS.[7] The recognition was in the form of a recommendation that "existing indigenous communication channels should be examined and developed;" and that this would "make it possible for an incorporation of such channels into the overall communication plan."

Means of communication are made to perform. They do not initiate performance by themselves. Nevertheless, as noted in Chapter 6 by Bourgault, the ritual (of the indigenous communication genre) is one type of communication that classically represents the problematics of isolating means from the social dynamics of which it forms part. The difficulty arises out of the fact that ICS exist within the holistic nature of African social systems. However, someone or people have to initiate the ritual activity itself.

the venue-oriented village communication described by Ansu-Kyeremeh (1997) whereby people gather at certain spots to exchange ideas, share knowledge and debate issues.

The new media are also "de-massified," as proposed by Freire (1970), to the degree that a special message can be exchanged with each individual in a large audience. Such capacity to individualize within the mass likens the new media to face-to-face interpersonal communication, except that they are not face-to-face.

The new communication technologies are also asychronous; meaning they have the capability for sending or receiving a message at a time convenient for an individual. The participants do not need to be in communication at the same time. Recording a live programme with the VCR for later (delayed) viewing is a typical example.

Whatever implications ICS might have for ICT (in their convergence as the Internet and other formats) there are the possibilities of improving convergence by closely examining the principles underpinning the dynamics of the ICS.

This book connects the past to the present; while projecting the close of the 20th century to the 21st century with its predicted communication technology convergence in which roles for the ICS are still expected.

Since the completion of this book, the Internet has made an impact in Ghana. The social and cultural ramification of some Internet-propelled communication events call for an examination of possible linkages between ICS and the Internet. Such an examination would serve two important uses. First, the analysis could contribute to further understanding of the limits and possibilities of the ICS in a sophisticated information and communication technology (ICT) world of communication convergence. In addition, it would provide an insight into whether indigenous communication has a place in the convergence.

In the words of Bates "all media have their strengths and weaknesses, and what is a strength in one medium is often a weakness in another."[2] In the case of ICS, besides their many utilitarian characteristics described in Chapter 1, Asamoa-Anane underscored their capacity for trade promotion.[3] The contemporary communicative power of the indigenous communication instrument, *dawuro,* for instance, was recently the source of struggle over its control between traditional authorities and the modern political administrative unit, the District Assembly.[4]

However, beyond the capacities of various forms of communication, the question is often asked if technological sophistication is coterminous with the performance effectiveness of a means of communication. Scholars

That the Internet shares the interactive characteristics of many ICS means that perhaps, this is an area for fusion between it and the local communication formats. Networking similarly exhibits many of the characteristics of small group interactive communication attributes that are among the strengths of the ICS (Ansu-Kyeremeh 1997). Thus, also implicit in the Internet's networking capacity are the possibilities for media "de-massification" which Freire (1979) for example, finds necessary for successful communication in communal situations.

Actually, the proposals for itnegration of ICS and modern media provide the example of the "adaptation" of the former to the format of the radio.[13] A Ghana national communication and information policy document also recommended the reexamination and development of indigenous communication channels for possible incorporation into the overall communication plan.[14]

The frontier-busting capacity of the Internet is now well-known. Its potential to affect culture is almost without limits. Paradoxically, many believe this is one quality that provides possibilities for countering its power of acculturation. All a nation needs to do is to input its own culture. (Actually, some believe it is the one tool yet for nonwestern cultures to make inroads into the western-dominated world culture.) In order words, like it or not, no culture can be insulated or even innoculated against the cultural intrusion, incursion and impact of the Internet. What cultures such as Africa's, therefore, need to do to resist Internet acculturation is to create the desired images on the Internet. Herlan, from this position, recommends "critical assessments to determine information for cutlural production and exportation" through the web.[15]

However, such technologically deterministic[16] approaches to the Internet raise the old questions of ICT consumption-deepening. Since the hardware and software for Internet operations are produced elsewhere the technology will have to be acquired. The Internet in this sense negatively helps sustain the canker of technology-driven AIDS (acquired imports dependency syndrome). Such costs outweigh any possible benefits accruing from information the net provides about AIDS the disease. On the other hand, opportunities for software development could be an incentive for consumption-oriented cultures to switch to software production to increase their input into the Internet. By this, they will not only be enhancing their images but will also be increasing their economic benefits from the Internet.

Internet itself has come to be associated with a communication chain. The chain is effected through cross-sourcing, intrainternet or internet-internet, internet-extrainternet-internet, input-output directional flows of

The McLuhanite doctrine of "medium is the message" may challenge the ability of humans to control the messages channelled through the media. Nevertheless, Von Hentig actually argues that the character of the medium transcends the role of a mere channel which passively conveys knowledge from source to destination.[8] In addition, it is not without reason to argue that it is only when humans fail to direct the medium and thus lose control over it that its initiating capacity could be activated.

Secondly, new forms of communication developed with technological superiority rarely completely edge out or render totally obsolete earlier ones.

Thirdly, media impact studies often find that the individual's decision-making is almost always influenced more by interpersonal communication than by mass communication.[9] In other words, whereas the mass media may influence what the individual may want to think about, the individual's decision to act tends to be influenced more by opinions shared with close relatives, friends and colleagues.

Communication is culture and culture is communication. That is to say the most effective communication occurs in circumstances where the symbols and the social dynamics of the host culture are utilized.

The challenge of Internet as ICT, from the discussion so far, is how best to use it effectively for development (as in research and website information on practical skills and techniques or professional knowledge) while at the same timee minimizing its inherent negative effects on the sociocultural and physical milieu (such as pornography or environmentally harmful website information). The Internet's interactive characteristic suggests that perhaps, more than any other technologically mediated communication system, it provides the greatest potential for participatory communication. And by its easy integration (even if in principle) with the indigenous systems, it stands to generate desired results with minimal negative impact.

In fact, combining technologically mediated communication with the largely interpersonal indigenous communication for greater effectiveness of communication in support of development is the centrepiece of the indigenization framework by which technological media will be selected for application based on their capacity to fit into the indigenous patterns (Ansu-Kyeremeh 1997). A similar proposition regarding indigenous knowledge in general is made by Hountondji[10] and Brohman.[11] Indeed, Fraser and Restrepo-Estrada[12] cite an integrated form of communication which fuses interpersonal, mass media, group media and traditional or folk media as a better option for social mobilization.

the chain. In this sense, one needs to be reminded that indigenous communication in its interpersonal and small group forms is known to be the ultimate decider of mass communication effects.

In many ways, current observable representation of ICS in the multi-media feature of the Internet, seems more symbolic than substantive. For example, Akan nomenclature (including *Okyeame* or spokesperson, *Dawuro* or gong gong and *Atumpan* or talking drums) have been adopted for various listeners by which messages are forwarded or redistributed.

Indirect links of the ICS to the Internet are also facilitated by the other media feeding into the communciation chain. Recently, the radio station *Radio GAR* is reported to have tapped the *dawuro* concept to rally/ mobilize listeners to its live broadcast of a traditional event, the Homowo festival of the Ga people.[18] From the town of Amasaman, linked to the main studio in Accra using outside broadcast facilities, the event ended up at the website of *Radio GAR* on the Internet.

There is also the talk of using *adinkra* ideographs and drum language as gateway symbols and cues. Furthermore, local languages are found suitable for application in certain types of messaging.

Without doubt, one of the most promising possibilities for combined Internet-indigenous media operations is in pictorial representations. The simplicity of *adinkra* ideographs can be extended to a new approach to literacy where illiterate beginner clientele of the Internet could use pictures to communicate through that medium. In his study of modes of communication among some rural Ghanaians, Ansu-Kyeremeh (1989) found that posters attracted more attention than any other media form. It is thus not impossible to design software, using pictorial resprentations, that will enable e-mail and other Internet messages to be shared between and among illiterates. Stratified statuses, such as the Akan *firatamni* (never-been-to-school) as against *krakye/awura* (schooled) have created a tendency whereby the former aspire to achieve the latter. There is the story of a woman whose desire for the *awura* status (of which the ability to read the newspaper is an important indicator) was so strong she held a paper upside down pretending to be buried in reading its contents. A "piture-perfect," illiterate user-friendly Internet would attract such eager learners.

Indeed, the Internet may share more of the characteristics of ICS than the other technologically mediated communication systems that preceded it. The Internet's multimedia capacity, for example, lends itself easily to incorporating the announcement role of the *dawuro* and *nkwong* (both gongs) as well as the face-to-face patterns of communication. A close information. By these modes, messages generated on the internet are spread through other means including television, radio, newspapers, magazines and finally interpersonal communication. As indicated in Chapter 1, much of interpersonal communication belongs to the family of the ICS.

Three recent national media and communication related incidents illustrate the Internet-driven communication chain. In April 1998, a widely circulating newspaper published nude pictures (with blocked faces) on its front page.[17] The publication instantaneously drew criticism from politicians, religious leaders (and on a wider scale) from listeners to various radio phone-in programmes. Upon investigation by the police, the following chain was established: Internet website — radio show — newspaper — individual word-of-mouth — interpersonal communication.

Not long after, news of the sudden death of the former head of Nigeria's military government, General Sanni Abacha, was all over the place hours before local radio stations, television and newspapers reported it. This time, the chain was from internet website/e-mail to other forms of communication including word-of-mouth, radio, television and newspapers. Hardly had people recovered from the shock of the Abacha death when the death of Abacha's political foe whom he had imprisoned, Chief Moshood Abiola, was announced through a communication chain similar to the one involving the Abacha death.

Internet as the obituary is now becoming common as news of the death a Ghanaian deputy minister in London was broken through the internet/e-mail nexus. On some of the above foreign-originating sources of the Internet-initiated communication chains, foreign media, especially, BBC radio has featured somewhere in the chain. In its role as prime source of the chains, the Internet seems to be assuming the role of the virtual rumuor machine. Rumours no doubt thrive best on word-of-mouth, a key form of indigenous communication which depends on authoritativeness for credibility. Each means of communication in the chain has its own multiplier effect and thus speeds up the spreading of news. In times gone by, the effect of such rumours would be confined to small sections of populations. These days, with Internet technology, it is a massive spread with lightening speed.

It needs to be recognized that in the African situation of the likelihood of higher illiteracy and low access to Internet technology, it is the rumour chain that will link the majority to the Internet as a source of information. That is to say, whatever impact Internet messaging is likely to have on the ordinary people is likely to depend on the role ICS will be made to play in or body language including strict codes involving the use of the left hand.

(h) Status may be a factor in communication.[19] The social environment contributes to communication. Various institutions lend support or even prescribe modes of communication. To that extent, the effectiveness of communication hinges on meeting the requirements of these institutions. The socio-political and cultural environment includes centripetal political organization, small-groups and face-to-face contacts all of which in their own way affect communication. Indigenous social organization also tends to be steeped in free movement, free association, free assembly and free expression traditions.

Attributes of the Internet

With the itneractive principles underpinning indigenous communication may be matched the capabilities of the Internet which derive from similar attributes. Various feedback-enabling Internet operational formats include the electronic mail (e-mail), the home page and the world wide web which elicit delayed feedback. There is also computer talk or chatroom which operates like a telephone with written instead of spoken words. Computer talk or chatroom, unlike the other Internet operations allows instantaneous feedback.

Sound and visual effects, although not as "natural" as in indigenous face-to-face interpersonal communication, are also characteristics of the Internet. Internet communication is almost unrestricted thereby emphasising casualness, less encumbering protocol, and no censorship.[20]

Convergence and Divergence

The above principles of indigenous communication matched against the capabilities of the Internet suggest various areas of convergence and divergence. For one, the nature and importance of timing in indigenous communication and of the communicative act is not dissimilar to peak and lean period uses of the Internet.

Children, women and others who may be disadvantaged by the dynamics of the social handicaps of indigenous communication could be empowered by the proactive universal access (barring literacy, economic and technological handicaps') characteristic of the Internet. The anonymity look at the principles that govern ICS could throw more light on their compatibility with the Internet format.

Principles of Indigenous Communication

From various discussions that centre around the nature of ICS, one may draw certain principles that underline the operations of those systems. They include:

(a) Communication is a shared experience.

(b) Following from the sharing capacity and symptomatic of Freire's (1979) contention that there can be no communication without feedback, communication is basically a two-way process, and feedback is an essential element in communication.

(c) Both one-on-one and group communication are allowed; although there are distinguishing aspects between the two modes. Dyadic communication does take a different form from group-oriented communication. In one-on-one instructional situations (e.g. of adult-child), silence may be interpreted as insult whereas things would be different in a group situation. It would be regarded as "attentiveness" in that situation.

(d) Anonymity (empowering and emboldening e.g. on radio phone-ins) is less an attribute.

(e) There are rules prescribed by custom to regulate communication. For example, time is of essence to communication. Communication tends to be more intensive on work-free days (*afofida*). Meetings and important discussions are held on such days or evenings. In fact, chances for daytime conversation during working days are limited. However, week-day evenings are not just the usual times for conversation but they are also periods for *anansesem* (storytelling).

(f) There are also protocols including various salutations.

(g) Language use is guided by standards as they affect courtesies, titles, actions, gesticulations, facial expressions and movement associated with the Internet could be simililarly empowering for such communication disadvantaged groups. As suggested earlier, anonymity is indeed already) having a tremendous impact on access to radio phone-in programmes. Thus, the Internet could empower where indigenous communication disenfranchises. One needs to add, though, that the mechanics of indigenous communication, provide room or make it possible for individuals to develop *ka na wu* (speak the truth and die) or *pae mu ka* (pull no punches) attitudes and reputations in defiance of the lack of anonymity.

Summary

From the foregoing discussion, there are possibilities for incorporation of indigenous communication in the convergence of different methods of communication that characterize the operation of the Internet. The communication enabling principles underlining the former may even inspire further improvements in the capacity of the Internet. However, one is under no illusion that all this is possible and that the cultural eroding power of the Internet would not outweigh any benefits accruing from it to societies dominated by indigenous communication formats. Perhaps the Internet would even undermine the fairly accessible indigenous communication by heightening the domination of communication by the elite using their economic and literacy power.

NOTES

1. See Ghana, Ministry of Communications. Draft Communication Policy Document (for National Conference on Communications Policy), 1998.

2. A. W. Bates, Using television in distance education. Paper presented at tbe 13th World Conference of the International Council for Distance Education, Melbourne, Australia, August 13–20, 1985.

3. Kofi Asamoa-Anane, Traditional channels and modes of communication Ashanti. Graduate Diploma Project. School of Communication Studies, University of Ghana, Legon, 1977.

4. "Mr. Stephen Kwadwo Ofori, a Community Tribunal chairman reminded assembly members [of the Dangme East District Assembly at Ada-Foah] that the beating of the 'gong gong' in towns and villages is a traditional preserve of chiefs and it is wrong for them (assemblymen) to go outside their duties to perform that function," said a *Ghanaian Times* (Friday, August 7, 1998, front page) story.

5.   Article 39 of the *Constitution of the Republic of Ghana, 1992.*

6.   Ghana. Ministry of education and Culture. Culture Division, *Cultural policy in Ghana.* Paris: Unesco Press, 1975.

7.   Ghana Ministry of Information, Report on the National Seminar on Communication and Information Policy for Ghana, July 6–10, 1992. (Mimeo)

8.   H. Von Hentig. What are the "media"? What is their rational function and where does it end? *Education,* 1986, 34: 43–51.

9.   K. Andreas Fuglesang, Beans in a bowl: Observations on communication and adult education in developing countries. *Development Communication Report,* 1979, 27: 8–10, 28: 8–9, who believes "nothing can beat person-to-person contact" supports this assertion with the view of Jackson and Moelino that "the type of sustained behavioral change that is needed if social change is to emerge does not result from use of mass media alone, but from learning situations in which technological media are combined with person-to-person communication and group interaction," p.9.

10. Paulin J. Hountondji, African cultures and globalisation: a call to resistance, *D+C.* 6, November/December, 1997, pp. 24–26.

11. John Brohman, *Popular development: Rethinking the theory and practice of development.* Oxford: Blackwell, 1996.

12. Fraser, Colin and Restrepo-Estrada. *Communicating for development.* London: I.B. Tauris Publishers, 1998.

13. Ghana, *Cultural Policy.* Paris: Unesco, 1975.

14. Ghana, Report on the National Seminar on Communication and Information Policy, 1992, *ibid.*

15. Pharra J. Herlan, Information technology in Ghana: Content development. Paper presented at One Day Information and Communications Technologies (ICT) Forum, Accra, Ministry of Communications, August 12, 1998.

16. Technological determinism may be interpreted according to the view of H. Mowlana (Communication and development: everyone's problem. In C. Okigbo (ed.) *Media and Sustainable Development,* Nairobi: ACCE, 1995, p.45) as that "the key to increased productivity is technological innovation, no matter who benefits or who is banned."

17. *Graphic Showbiz,* April 14–22, 1998, p.l.

18. As told by Cris Tackie Acting Director of Radio at GBC at the "One Day Forum for tee Electronic Media (Broadcasting, Video/Film)," Ministry of Communications, Accra, Tuesday, September 1, 1998.

19. Observation by Kwaku Owusu-Yeboah, Traditional communication in the Akan society: A case study of the Techiman area. Graduate Diploma Project School of Communication Studies, University of Ghana, Legon, 1988.

20. The spread of socially undesirable messages such as pornography and racism through the Internet is causing some governments to seek controls of Internet use. For example, the British government (according to BBC "World News," Tuesday, April 28, 1998) was seeking to obtain the key to encripted; messages to enable the police decode such Internet messaces. The *Daily Herald* News," Tuesday, April) 28, 1998) was seeking to obtain the key to encripted messages to enable the police decode such Internet messages. The *Daily Herald Tribune* (January 2, 1998) also noted China clamped sweeping new controls on the Internet.

Chapter 16

"DIGITIZED STORYTELLING"[1]

Kwasi Ansu-Kyeremeh

Introduction

This chapter explores the proposition that the more one moves away from storytelling as known in its original face-to-face format, the greater the relevance of the basics of its communicative role. Plain old interactive storytelling continues to thrive even in the face of radical technological advancement in the format in which it is communicated. This is so, perhaps, because people still want to listen to, imagine, dream, visualise, hope, laugh and relax with stories. Audiences also want to be able to sympathise and empathise with characters in stories as they feel the roller-coaster effect of exciting episodes. It seems then, that with storytelling, the more things change, the more they remain the same. (This old saying is true even in the technological transformation of communication.) Another characterisation, though, would be that the basics or the generic attributes of the storytelling act are never lost no matter the level of the technological sophistication with which the narration is performed.

The structure of a story having a beginning, a middle and an end (explained in detail later using the example of *mmoguo* or interjection), for example, seems to apply to storytelling even in the most technologically sophisticated format. Take virtual storytelling which is the subject of this Chapter. It utilises the basics of communication using the medium of storytelling as does indigenous communication systems (ICS) such as the Akan *anansesem* (storytelling). Perhaps unlike other forms of ICS, storytelling is as much alive today among Internet users, especially the cyber community (small group) type, as it is among the Burundian, Ghanaian, Nigerian, Ugandan or Zimbabwean rural community.

Storytelling as Communication

The activity of messages flowing between and among a small group of people comprising a storyteller and his/her audience is communicative in that it, by its ingredients of the exercise, represents encoding (sending) and decoding (receiving) signs and symbols. Thus, it has the dynamics of communication at play. As indicated later, the flow could be vertical or audience, then, elasticity in audience composition means the small group is a large group.

Storytelling through Newspaper Communication

Evidence of continuity in storytelling as a communicative act in the Ghanaian socio-cultural context is in the incorporation of short stories in the contents of the newspaper when it was introduced as a tool of communication. The "change" simply represents transformation of the oral format into a written format. *The Western Echo*, one of the early Ghanaian newspapers which was published in 1885 by James Hutton Brew, carried a regular story column called "Fanti Folk Tales."[3] In addition, the *Echo* published "Marita or Folly of Love," a serialised story with the byline "By a Native." There is no study that documents the extent to which the several newspapers that have appeared, appeared only to disappear, or are in existence, carry short stories. Nevertheless, short stories have since the *Echo* days continued to be a major feature of some newspapers. Actually, one of the longest lasting weeklies, *The Mirror* (previously the *Sunday Mirror*), regularly publishes short stories. The continued use of short stories to fill newspaper space indicates reader interest in them.

Storytelling on Radio

If the Ghanaian newspaper found folk tales as useful and irresistible content material, storytelling proved logical in programming on radio, whose oral format made it a natural channel for storytelling. Like newspaper content, one of the earliest programmes broadcast upon the introduction of radio broadcasting into the then Gold Coast (Ghana) on July 31, 1935 was storytelling. For decades, the signature tune for storytelling time on *GBC Radio* has been resonating in the air drawing attention to its storytelling programme. The tune is sung in Akan Mfantse in the following words:

> *Kwaku*[4] *de onsuro oo, Kwaku de onnsuro oo,*
> *Kwaku e, Kwaku e,*
> *Awo a wo de biribi nkotum wo da*
> *Kwaku na dabew ahia wo yi;*
> *Ene a Kwaku ee, Okyena a Kwaku ee;*
> *Awo na dabew ehia wo yi.*

The programme is produced by the Rural[5] Broadcasting unit for the horizontal. Factually, the presence of such basic characteristics of the communicative act, including encoding or composing the story by its originator and telling or sharing it with others as decoders or recipients (who are sometimes participators by providing feedback) establishes a communication construct.

This is all strengthened by the need for communication to have goals or objectives that are to be met in its effects. Intended effects are fashioned out as beneficial to communicator and communicatee. In situations of miscommunication or communication gap, though, intended benefits become costs. A recent desire to "revive storytelling in libraries and popularize it in schools and communities"[2] was justified as "necessary because the nation's rich tradition of folk stories were in danger of disappearing as a result of western education and modern sources of entertainment like television, radio and cinema." Also "through storytelling, societal values and norms 'are inculcated in the *hearers* [emphasis mine], be they children or adults and are made aware of good moral and social behaviour, made proud of their heritage, learn to understand their culture and develop a sense of belonging.'"[3]

A Small Group Activity

Indigenous storytelling seems cut out as a small group activity. The small group refers to patterns of face-to-face interaction in which the individuals involved are few. One is looking at fewer than ten members and rarely more than fifteen. As so often stated, the venue is often by the fireside when children warm up to escape from the cold of the nights.

The BTF type which is a combination of ICS and television formats has audience implications. For example, there is a clear distinction between studio audience and the viewing audience. As actors in the performance-oriented in-studio communication, the studio audience can actively participate in the storytelling event. The viewer audience can, however, only vicariously associate with the activity. Its membership is without choice passive listeners and spectators. Even where a live post-viewing discussion programme is organised, any input or feedback is best delayed.

Another dimension of audience size is the possibilities for expanding the audience base. The multiplier effect of technology-based communication such as recording TV programmes for VCR play, enables audiences who would have missed the original telecast to still view the programme. In addition, those who are interested in repeated viewing are enabled by this mode. It is this "saved" and repeatable elements that bring the DS format to the BTF in its technological aspects. In the context of the storytelling

Radio One Network of the Corporation. Storytelling is classified among the Corporation's entertainment (rather than education or information) programme genre.

Another storytelling programme (which uses English as medium), "Once-Upon-a-Time," is produced by the Radio Schools Department for its Schools Broadcast. It is a "story telling programme for [primary] classes five and six" (GBC, *50 Years of Broadcasting in Ghana*, 1985, p.21). Rather than classify this children's programme as purely of the entertainment genre, by Ngabirano's earlier analysis, it would belong to the "edutainment" category of programmes because of its built-in educational content of teaching children morality as indicated in the assumption by the GLB that storytelling in the library contributes to children's knowledge build-up. Until some five years ago, the Corporation's Ghanaian language channel, *Radio One*, broadcast a weekly Akan storytelling programme "Kodzi (*Anansesem*)." It, indeed, still broadcasts a weekly (Tuesdays) fifteen-minute storytelling programme, "Salima Saha," in the Dagbani[6] language.

Among the popular culture conscious and "hip-life music loving youth, though, what seems more appealing is less storytelling and more of the joke-telling programme format known locally as "Toli Time." It is a kind of stand-up comedy on air. Brief stories are told in jocular format. It is widely adopted by FM stations as a purely entertainment programme with little or no educational value.

Audio-Visual Technology and Storytelling

To some level of transformation, indigenous storytelling has inspired radio and television programme production in cases genres such as the telenovella or even the sitcom. In Ghana, though, until the introduction of cyber storytelling, the most technologically sophisticated form of storytelling had been the *GTV* enactment of storytelling time with its programme "By the Fireside" (BTF). The story is unlikely to have been different in other African societal contexts. As indicated above, radio had run story time since its beginnings. With the *GTV* BTF programme, once-a-week and for one half hour, an adult tells children who gather around him/her stories in an enactment that captures the elements of the age-old indigenous storytelling act. These include the opening, the story itself and "*mmoguo.*" BTF is a recorded television programme.[7]

The village stratified communication (by age as reflected in music and dance and sometimes by gender (Ansu-Kyeremeh 1997) is replicated in age as children are restricted from accessing certain adult websites (just like they could not participate in certain adult converstaions at village meeting spots.)[8] New communities are also created through websites or such programmes as the German model of "how to use the net" lessons for seniors "Project 50+," in which a youngster teaches pensioners how to use the Internet (*DWTV*, Tuesday, December 31, 2002). Apparently, communication technology was far from its current computer applications in the youthful years of these senior citizens. As stated in the previous chapter, even more stratified is the digital divide.

In the early versions of interactive computer messaging such as computer talk, moderated conversations occurred in typewritten alphabetic characters as differentiated from the spoken telephone (which is incidentally needed for the modem to be functional) word. Cyber storytelling has, however, taken the act one step further. From the delayed feedback talk e-mail messaging, including the multiplier effect of listserv and message forwarding, interactive computer communication has moved into exchanges in which the modem utilises the telephone to enable one computer to communicate with another.

Participatory Communication

Perhaps the most attractive attribute of storytelling as means of communication for rural development, for example, is its participatory nature. Characterisation of the storytelling audience as "hearers" as was stated earlier, hides the two-way participatory nature of storytelling as communication. The notion of "hearers" implies passive and ineffective listening and one-way communication as articulated by (Ross and Dewdney 1989). Contrary to this view, active listening is a key activity in storytelling. Even in the adult-child BTF format, questions and comments are facilitated by the two-way communication (exemplified in *mmuguo* and comments/ questions) between storyteller and his/her audience. The importance attached to involvement through feedback mechanisms of the audience in the communication process has been identified by experts such as Freire[9] as indispensable if communication is to take place at all.

Mmoguo, Instrument of Interactivity and Participatory Communication

A typical example of a participatory enabling device in storytelling is the *mmoguo* (which literally means rejection or casting aside). *Mmoguo* is a musical device of occasional song and dance that is employed to punctuate the storytelling process. It is designed to enliven the process by breaking the possible monotony of the storyteller's continuous narration or, simply, to spice the act. The four types of listening identified by Lynne Meade (*http://lynn.meade.tripod.com/id26m.htm* [viewed January 27, 2003), namely: listening for information, critical listening, empathetic listeningand listening for enjoyment are all present to help construct the *mmoguo* and its rendition.

An analysis of the *mmoguo* segments of the BTF storytelling as a communication format points to a song of a lead solo followed by a chorus response by everyone present. *Mmoguo* also serves as an enhancement or critical feedback device in the storytelling activity. However, as an enhancement, it seems incongruous given that the literal meaning of the term is rejection or casting aside. This interpretation notwithstanding, often, a member of the audience is able to express dissatisfaction or boredom with the storyteller by interjecting the narration with an *"mmoguo."* *"Mmoguo"* itself connotes a kind of protestation. A particular *mmoguo*, is direct in its criticism through confrontation and demand for excitement in the storytelling act. The lyrics are: *"Anansesem ye asisie to no yie."* This translates into: "Storytelling is a fabrication or construction exercise so get your elements, logic and flow right."

Such faulting of or during the storytelling process seems to find expression in the notion of "false construction" which, as articulated by Parmar (1975: 70), has dangerous implications for a story's socio-cultural impact. Where storytelling is for the purpose of education (of the young), the likely negative impact is even greater. What *mmoguo* does as a corrective feedback is that the storyteller is instantaneously reminded about the displeasure of the audience, the unentertaining quality of the story (or its telling or a combination of the two) or the unacceptable values embedded in the content.

The structure of storytelling is also effectively explained in the dynamics of the *mmoguo*. An *mmoguo* introduces the storytelling activity. It is used by audience members to formatively assess the quality of the story. And then certain types of *mmoguo* are used to close the story.

An identification of the two types of the Akan *anansesem* (storytelling) format is useful for the purpose of differentiating or isolating the various dimensions of the *mmoguo* act. They are the adult-child vertical communication session and the peer horizontal communication session. The adult-child session involves an adult (an elder usually) telling stories to children gathered at a venue such as the fireside. Adaptations of this format to the technological media includes *Radio Ghana's* Akan *Anansesem* programme and its Dagbani version, the "Once-Upon-a-Time" GBC English Schools Broadcast programme and *GTV*'s current "By-the-Fireside." Feedback during these storytelling sessions may include questions such as: "Grandpa, why did Ananse have to cheat?" Protests from the audience such as "Aaa, that's not fair" are also not uncommon.

The peer storytelling sessions involve age-grade groups which gather at certain venues for *anansesem*. Rather than a singular storyteller, different individuals take turns in horizontal communication flow to tell stories. The audience member and the storyteller are coterminus. That is to say, members of the audience are also the storytellers; they take turns to tell stories. Comments and reactions are more direct and less restraining in language use than the adult-child format.

Descriptive structural analysis of the storytelling act (in either format) reveals a beginning in which the storyteller calls the audience's attention with "*Abra, abra abra aa!*" To this the audience responds in chorus: "*Yoo, yoo!*" Indeed, this is the approach adopted for the BTF. One is unable to interpret or decipher the literal meanings of the two expressions by providing English equivalents. However, the storyteller seems to establish the readiness of the audience (for example whether they are all ears) with the first while the second is the audience's express indication of listening readiness. From this perspective, one may reasonable equate a combination of the two expressions to the English: "Once-upon-a-time! Time, time!"

The above beginning of the storytelling act may be preceded by what may be considered part of the general preparations towards the act. In the peer format, for example, the storytelling proceedings would open with the *mmuguo*: "*Si sen o si sen o? Esi wo ara*" ("Where is the point, where is the point? They say it's your point;" or more appropriately "Whose turn, whose turn; It's your turn").

Other types of *mmoguo* are sung in the course of the storytelling process. The lyrics of these intervening *mmoguo* are usually determined by the particular story being told. There is, however, a specific *mmuguo* which essentially criticizes the storyteller and his/her storytelling style and/or abilities. It is "*Anansesem ye nsisie, to no yie!*" which was described earlier. It is generic or it is sung at all kinds of peer storytelling sessions. Its application is also universal — that is, it is sung irrespective of the type of story being told.

Then there is the ("*requiem*") *mmoguo* of closure which also applies in all peer storytelling situations. With this *mmoguo*, an audience member in a lead role begins with the solo: "*Agya Kwasi mmoguo!*" and the rest of the audience joins with the chorus: "*Yaa mmoguo!*" The singing goes on until the storyteller prematurely terminates the story. This *mmoguo* is designed to stop, and indeed, does abort a story in progress. Its rendition is an indication that the narration is judged by consensus among the audience to be boring or repetitive "beyond repair." Essentially, it translates into the inability of the storyteller to address, through modifications, the audience's dissatisfaction with the storytelling act as expressed through the warning *mmoguo*. An example of this type of *mmoguo* is: "*Anansesem ye asisie to no yie.*" Thus, "*Agya Kwasi …*" is most certain to always follow "*… to no yie.*"

A normal story ending will have the storyteller (in both adult-child and peer formats) concluding with the self-critical review remarks: "*M'anansesem a metoo ye yi, se eye de o, se enye de o, ebi nnko na ebi mmra.*" In the peer storytelling format in which members of the audience take turns to tell stories, though, the storyteller designates a specific person to continue the session with the next story, instead of ending with "*… se eye de o …*," he/she adds "*Mekyekre me soa wo;*" to indicate who's next.

Another significance of *mmoguo* as a song interjected into a storytelling act is the tendency among some Ghanaian musicians to use the words of various *anansesem* as lyrics for their songs. A number of the lyrics of (especially Akan) recorded music are short *ananse* stories adapted to musical scores. The words are borrowed from *ananse* stories. One particular popular musician, Nwomtofohene Nana Kwame Ampadu is acknowledged as an effective storyteller with his lyrics.

"Digitized" Storytelling

Pursued by organizations including the Digital Clubhouse Network (*http://www.storycenter.org/clubhouse.html*),[10] digitized storytelling utilises the technology of the web to construct stories. Through workshops, and in some cases semester or summer courses, learners are taken through the mechanics and processes of writing stories using the digital media. Participants in courses are taught "storytelling and digital media skills in a collaborative setting." This mode of storytelling assumes the mode of small group communication without the latter's natural nonverbal attributes and participatory enabling capacities.

In a sense, the Clubhouse is a "glocalized" form of *anansesem*. Glocalization[10] is an emerging term used to describe the introduction of digital communication and its functioning in the context of the application of technology-based mass media of communications within the "traditional" locality.[10] The availability of "digitized storytelling" is a recent development.

In Ghana, Internet access is severely limited to the paltry percent of the population who can and are able to use it.[10] Add the fact that only literate persons can access it in a country with a literacy rate of 49 percent, and the limitation is further diminished. With the Internet, cyber storytelling fans are enabled to visit the "Digital Clubhouse," the virtual cyber storytelling site.

Stories are, however, mostly likely to be written and pasted at a website. It accordingly assumes the status and form of the book. At the opposite end of the in-person presence of both storyteller and audience, the author is out-of-sight. All possibilities of instantaneous feedback are, thus, closed. The added nonverbal influence on the story told face-to-face is equally absent.

ELEMENT	ICS MODE	DIGITAL MODE
Setting/Social Context	Family/Community	Cyber Community
Venue	Fireside	Website
Direction	One/Two-Way	One-Way (Interjection not possible)
Actors	Adult to Children	Among Peers
Mode	Oral Interactive	Typewritten
Medium	Word-of-mouth	Electronic transm
Format	Narration + *Mmoguo*	Writing
Feedback	Instantaneous	Not necessarily Instant or even / Delayed
Type	Verbal/Non-verbal	Verbal only
Required Skills	Oral expression	Internet literacy
Audience	Small and Specific	Defined by Website Use
Contact	Face-to-face	Impersonal
Word Form	Spoken Word	Written Word

Fig. 16.1: Elements of "Storytelling" as Communication

Elements of "Digitized Storytelling" Communication

In general, the communication activity involves the interface of elements set out in the above schema (Fig. 16.1). The schema attempts to separate the ICS mode of communication from the digital mode. The elements include the socio-cultural setting in which the communication occurs or the social unit of its occurrence. One is also looking at whether the communication is venue-based, games-oriented, performance-oriented or event-related (Ansu-Kyeremeh 1997). The direction of communication, the actors involved, mode, medium utilised and the format constitute other elements of communication. A capacity to generate feedback, type of communication, the audience, nature of interactivity or contact, word form and required skills for participation are also important elements of communication.

From the website *http://www.storycenter.org/clubhouse.html* (viewed January 29, 2003), one learns that digitized storytelling (DS) exhibits many of the characteristics of the storytelling of old. It draws on the notion of the community and its attributes to teach "storytelling and digital skills in a collaborative setting." Ordinary stories are transformed into digital media. A group of students developed "their own story web site called The Story Vault." Attempts are made to inject humanness into the mechanical state of technology.

The above schema actually sets out the commonalities and differences between the ICS and digitized formats of storytelling. The "clubhouse" characterization reflects a "chatroom" pedigree. Common characteristics between the two modes include the fact that stories are told among the cyber community of individuals devoted to the act. The stratified status of an adult storyteller and children audience of the by-the-fireside (BTF) format is replaced with peer communication in the DS format. Both utilise the small group communication type. Also, both provide vicarious rewards (as in satisfaction with performance of characters as well as in their just and fair treatment) and dissatisfaction (where sanctions are not applied or punishment is judged unfair).

Differences

For the adult audience of the web, jocular and relaxing effects are perhaps of greater essence than the morals inherent in the *ananse* story (as observed by McWilliam and Kwamena-Poh 1975 and Bentsi-Enchill 1971) told to children. The teletext activity of storytelling communication is impersonal unlike the face-to-face mode of BTF. A typical difference between the two formats is the total absence of interjection in the DS format. In digital storytelling, a storyteller needs to pause completely for the listener to type any form of response. With the oral BTF format, a listener can actually start an *mmoguo* as a signal for the storyteller to stop to allow the *mmoguo*'s full rendition.

Mmoguo is usually designed to enliven and spice the monotony associated with narration. BTF represents the reality with touch and feel while cyber setting is remote. DS feedback is likely to be more rational and analytical unlike children's "innocent" reactions in the BTF situation. BTF setting provides some element of natural affinity and closeness while the DS format engenders artificiality. The DS format has greater potential for creativity and elasticity than the BTF which is likely to experience recycling of key parts of particular stories. While the element of imagination, mirage etc. are apparent in the adult DS story, the child audience of the BTF story is likely to take things a little more seriously by attaching a degree of reality to it.

For viewers, who are not part of the studio audience, though, participation is nil and some of the nonverbal nuances and effects exposed to the studio audience might be lost on them. And unlike the studio audience, any association with the storytelling process, including satisfaction or dissatisfaction, is more likely to be vicarious.

Whereas the indigenous is restricted to the small group communication format, the digital cyber type uses the small group to tell the story which is accessed by a mass audience.

Glocalization and Maximization of Benefits

The globalized world is also a glocalized one. While the former represents ubiquity of technology, the latter underscores possible incongruities arising out of the application of technology in the "traditional" setting; especially the possible less beneficial and less effective domination of technology-based communication aimed at supporting development. Indeed, the implications arising out of the localized condition ought to be of concern to the development communicator. The consequences for the rural community in terms of limited access, higher costs, the prerequisite of literacy and other forms of communication disenablement in the rural "traditional" context, have already been discussed in previous chapters.

As is reiterated in the various chapters, in many ways, this issue forms the basis for the need to study indigenous communication systems which are embedded in the social and cultural dynamics of the communities in which communication occurs as a means of seeking the maximization of effects as propounded by Wilson. In the sense of the Digital Clubhouse, "glocalized" communication include sharing rather than disseminating or transmitting messages in the form of information and education.

Summary

An attempt has been made in this chapter to identify the characteristics of face-to-face storytelling as practised among members of rural communities and as is replicated in high-tech on-line storytelling on the Internet. The focus on storytelling as a form of communication that exists alongside several other forms, is only to reduce the general to the specific. The example, it is believed, underscores the need, possibilities and model construction effort to take into consideration existing indigenous communication systems in low technology rural communities. If encouraged, the expected desired results of effective communication, especially that which seeks to support development, would most likely be realised.

NOTES

1. *CNN* News (Tuesday, March 11, 2002) identified an Internet storytelling community known as the "Digital Clubhouse."

2. See *Ghanaian Times* (January 24, 2003, p.16) story "Library Board brings back story telling," with byline Jafaru, Musah Yahaya.

3. Telephone conversation (Friday, January 3, 2003) with Dr. Audrey Gadzekpo of the School of Communication Studies Studies, University of Ghana about her doctoral thesis "Women's Engagement in Gold Coast Print Culture, 1857–1957" (University of Birmingham, 2001).

4. Kwaku is the assigned Akan name to a male born on a Wednesday. It is unclear as to why the trickster Ananse is given this particular *akradin* (as described by Denis M. Warren, *The Akan of Ghana: An Overview of the Ethnographic Literature*. Accra: Pointer, 1986). Further description and explanations may be found in Ansu-Kyeremeh's examination of names and naming in the Bono (Akan group) context in his article "Communicating *Nominatum*: Some Aspects of Bono Personal Names," *Research Review*, 2000, 16(2): 19–33.

5. It is instructive to observe at this point that (irrespective of the medium through which it is channeled) communication with indigenous implications is virtually

always associated with the rural context and the medium of local languages. This is exemplified in the association of storytelling with Ghanaian language and rural broadcasting programming of the GBC.

6.   According to Mr. Amankwaa Ampofo of Radio One in a conversation at the studios on Friday, January 24, 2003.

7.   This compares to "The Story Lady" movie which is a one-time only creation, unless repeated as a programme or by VCR effect.

9.   Paulo Freire (1981) Extension or Communication. In *Education for Critical Consciousness*, New York: Continuum. Pp. 87–110.

10.   The theme for the 2003 Taipei Conference of the International Association for Media and Communication Research (IAMCR) is "Information Society and Glocalization: What's Next?"

11.   See Ansu-Kyeremeh (*Communication, Education and Development*, Accra: Ghana Universities Press, 1997) for discussion of "venue-based" communication in the village setting.

12.   In the view of Stephen Lax (as expressed in *Beyond the Horizon, Communications Technologies: Past, Present and Future*, Luton: University of Luton Press, 1997), that majority of the people who use the Internet utilise only its e-mail function.

REFERENCES

Abrams, T. (1984) Using traditional communication for development — report of a Papua New Guinea study and pilot project outline. In G. Wang and W. Dissanayake (eds.) *Continuity and Change in Communication Systems: An Asian perspective.* Norwood, N.J.: Ablex, pp. 65–80.

Aggarwala, Narinder (Spring, 1979) What is development news? *Journal of Communication*, 29(2): 18–33.

Agovi, K. E. (March, 1989a) Culture, the state, and the artiste, *Uhuru*, pp. 14–15.

Agovi, K. E. (September, 1989b) "The cultural officer as an administrator and manager." Paper presented at an orientation course for cultural workers, Specialist Training College, Winneba.

Akpabot, Samuel E. (1975) *Ibibio music in Nigerian Culture.* East Lomsing: Michigan State University Press.

Akpabot, Samuel F. (June 3, 1981) "Music as an instrument of social change in Africa." Paper presented at the Institute of African Studies Seminar Series, University of Ibadan, Ibadan.

Alexandre, Pierre (1972) *An introduction to languages in Africa.* London: Heinemann.

Amponsah, K. (1977) *Topics on West African traditional religion.* Accra: Adwinsa Publications.

Anim, G. T. (1976) Reconceptualizing the Role of the Press: The Case of Ghana. Thesis, Ph.D. (unpublished), University of Iowa, Iowa City.

Anon. (1975) *Getting the message across.* Paris: UNESCO.

Anon. (March 8, 1994) Hawaiian society before Western contact, *Spirit of Aloha*, pp. 8–11, 51–52.

Ansah, P. (1985) *Broadcasting and national development.* Accra: Ghana Broadcasting Corporation.

Ansah, P. (1986) Broadcasting and multilingualism. In E.G. Wedell (ed.) *Making Broadcasting Useful, the African Experience: The Development of Radio and Television in Africa in the 1980s.* Manchester: Manchester University press, pp. 47–65.

Ansu-Kyeremeh (1997) *Communication, Education and Development, 2nd Ed.* Accra: Ghana Universities Press.

Ansu-Kyeremeh, K. (1994) *Communication, education and development: exploring an African cultural setting.* Edmonton: EISA.

Ansu-Kyeremeh (1989) Communication in rural education: An exploratory study of the application of media to Ghanaian village education.

Thesis, Ph.D. (unpublished), La Trobe University.

Ansu-Kyeremeh (May, 1988) Contextual concepts for media-delivered village education, *Media Information Australia*, 48: 49–55.

Ansu-Kyeremeh, K. (1992) Cultural aspects of constraints on village education by radio, *Media, Culture & Society*, 14(1): 111–128.

Ansu-Kyeremeh, K. (1994) *Communication, education and development: Exploring an African cultural setting*. Edmonton: EISA.

Ansu-Kyeremeh (1997) *Communication, Education and Development* 2nd Edition. Accra: Ghana Universities Press.

Aranha, M. (1983) Cultural implications for instructional development, *Educational Media International*, 2: 17–21.

Arnold, Stephen (1990) Preface. In *Culture and Development in Africa*. Trenton, N.J.: Africa World Press.

Ashby, Eric (1964) *African universities and Western tradition*. Cambridge, Mass.: Harvard University Press.

Awa, Njoku E. (May 18–23, 1980) "Continuity and change in traditional and modern communication systems in Africa." Paper presented at the 30th Annual Conference of the International Communication Association, Acapulco, Mexico.

Awa, Njoku E. (June, 1989) Participation and indigenous knowledge in rural development, *Knowledge: Creation, Diffusion, Utilization*, 10(4): 304–316.

Bame, K. N. (1972) Contemporary comic plays in Ghana: A study in innovation and diffusion and the social functions of an art-form. Thesis, M.A. (unpublished), University of Western Ontario, London, Ontario.

Bame, K. N. (1975a) Comic plays in Ghana: An indigenous art form for rural social change, *Rural Africana*, 27: 25–42.

Bame, K. N. (1975b) A study of traditional and modern media for communicating family planning in Ghana (unpublished) Institute of African Studies, University of Ghana, Legon.

Bame, K. N. (1985) *Come to laugh: African traditional theatre in Ghana*. New York: Lilian Barber Press.

Bame, K. N. (1990) Some strategies of effective use of research information in Africa: Some Ghanaian and other African experiences. In S. Arnold and A. Nitecki (eds.) *Culture and Development in Africa*. Trenton, N.J.: Africa World Press. pp. 131–143.

Bellman, Beryl (1984) *The language of secrecy*. Trenton, N.J.: Rudgers University Press.

Bielenstein, D. (1978) Political and economic factors of a New International

Information Order. In D. Berwanger *Toward a New World Information Order: Consequences for Development Policy*. Bonn: Institute for International Relations, pp. 9–16.

Boafo, S. T. (1984) *"Wonsuom"* — a rural communication project in Ghana, *Development Communication Report*, 47: 3.

Boafo, K. (Ed.) (1989) *Communication and culture: African perspectives*. Nairobi: African Council on Communication Education.

Boal, A. (1979) *Theater of the oppressed*. (C.A. and M. Leal McBride, Trans.). New York: Urizen Books.

Bourgault, Louise M. (1987) Orality and literacy: An analysis of the Nigerian press, *World Communication,* 16(2): 211–235.

Bourgault, Louise M. (1989) Participant observation and the study of African broadcast organizations. In Sari Thomas (ed.) *Language, Thought, Media, and Culture.* Norwood, N.J.: pp. 342–355.

Bourgault, Louise M. (November, 1990) "Talking to people in the oral tradition: Ethnographic research for development communication." Paper delivered at the Annual Meeting of the African Studies Association. Baltimore, Maryland.

Brooke, Pamela (1989) "Adapting Village Drama to Radio." A Report presented to the Institute for International Research. Monrovia, Liberia, March, 1989. (Unpublished).

Busia, K. A. (1967) *Africa in search of democracy*. London: Routledge and Kegan Paul.

Byron, Martin and Ross Kidd (1977). The Performing Arts: Culture as a tool for development in Botswana. In *Botswana Notes and Records*, 10: 81–90.

Charles, E. (October, 1989) "Traditional communication and community development." Paper presented at the World Association for Christian Communication Congress, Manila.

Chernoff, John Miller (1979) *African rhythms and sensibility: Aesthetics and social action in African musical forms*. Chicago: University of Chicago Press.

Colletta, N. (1977) Folk culture and development: Cultural genocide or cultural revitalization? *Convergence*, X(2): 12–19.

Colletta, Nat J. (1980) Tradition for change: Indigenous sociocultural forms as basis for non-formal education and development. In Ross Kidd and Nat Colletta (eds.) *Tradition for Development, Indigenous Structures and Folk Media in Non-formal Education*. Berlin: pp. 79–97.

Csikszentmihalyi, Mihaly (1974) *Flow: Studies in enjoyment.* Chicago: University of Chicago PHS Report.

Dejenne, Alemneh (1980) *Non-formal education as a strategy to development: Comparative analysis of rural development projects.* Lanham: University Press of America.

Desai, Gurav (1990, April) Theatre as praxis: Discursive strategies in African Popular Theatre, *African Studies Review*, 33(1): 65–92.

Determeyer, H. (1992, September) *Het gebruik van acteurs en drama ij educatie en activiering in Western Province, Zambia.* Unpublished paper presented on a Seminar on Culture and Communication, organised at Clingendael by the Directorate General International Co-operation (DGIS), The Hague.

Diaz Bordenave, J. E. (1977) *Communication and rural development.* Paris: Unesco.

Dietz, B. W. and M. B. Olatunji (1965) *Musical instrument of Africa.* New York: John Day Company.

Dissanayake, W. (1977) New wine in old bottles: Can folk media convey modern messages? *Journal of Communication,* 27: 122–124.

Dissanayake, W. (1981) Development and communication: Four approaches, *Media Asia,* 8: 217–277.

Doob, L.W. (1961) *Communication in Africa: A search for boundaries.* New Haven: Yale University Press.

Douglas, Mary (1970) *Purity and danger.* Middlesex, England: Pelican.

Ekwueme, Laz E. N. (Juky 4–15, 1983) "Music, the Arts and Communication in Africa." Paper presented at the Postgraduate Seminar on Theories and Methodologies in the Sociology of Communication, Department of Mass Communication, University of Lagos.

Elliott, Philip and Peter Golding (1979) *Making the news.* London: Longman.

Ely, D. P. (1983) The use of educational communication in different cultures, *Educational Media International,* 1: 12–16.

Esen, A. J. A. (1982) *Ibibio profile: A psycho-literary projection.* Calabar: Paico Press and Books Limited.

Euba, Akin (March, 1982) Interview in *Toplife* magazine, 1: 12.

Fabian, Johannes (1990) *Power and performance: Ethnographic explorations through proverbial explorations in Shaba, Zaire.* Madison, Wisconsin: University of Wisconsin Press.

Finnegan, R. (1970) *Oral literature in Africa.* London: Clarendon Press.

Fiofori, F. O. (Spring, 1975) Traditional media, modern messages: A Nigerian study, *Rural Africana,* 27: 43–52.

Fiske, John (1982) *Introduction to communication studies*. London: Methuen.

Friere, Paulo (1970) *Pedagogy of the oppressed*. New York: Continuum.

Freire, P. (1974) *Pedagogy of the Oppressed*. New York: Seabury Press.

Freire, P. (1979) *Pedagogy of the Oppressed*. (M. Bergman Ramos, Trans.) New York: Seabury Press.

Fuglesang, A. (1979) Beans in a bowl: Observations on communication and adult education in developing countries, *Development Communication Report*, 27: 8–0.

Fuglesang, A. (1984) 'The myth of a people's ignorance', *Development Dialogue*, 1–2: 42–62.

Gay, John (1973) *Red dust on the green leaves*. Yarmouth, Maine: Intercultural Press.

Gecau, Kigami (1982) A Theatre for Development in Zimbabwe, *Africa Report*, 25(6): 24–26.

Geertz, Clifford (1973) *The Interpretation of Cultures*. New York: Basic Books.

Ghana (26 February, 1970) *Parliamentary Debates: Debate on the Newspaper Licensing (Repeal) Bill*, Second Series, Vol. 2, Col. 297.

Gill, D. H. (1984) *A framework for community development and extension education training in support of an integrated rural development programme in the Northern Region*. Tamale, Ghana: Northern Region Rural Integrated Programme.

Goody, Jack and Ian Watt (1968) The Consequences of Literacy. In Jack Goody ed. *Literacy in Traditional Societies*. Cambridge, England: Cambridge University Press, pp. 27–84.

Gortzak, H.-J. (1992) De obstakels voor een anti-AIDS campagne in Afrika, *De Volkskrant*, zaterdag 7 maart: 7.

Grant, J. P. (1988) *The state of the world's children*. Oxford University Press; Unicef.

Gyekye, K. (1987) *An essay on African philosophical thought: The Akan conceptual scheme*. Cambridge: Cambridge University Press.

Hachten, William A. (1971) *Muffled drums: The news media in Africa*. Ames, Iowa: Iowa State University Press.

Hall, Budd (1981) Participatory research: Popular knowledge & power: A personal reflection, *Convergence*, 14(3): 6–19.

Hall, S. (1977) Culture, the media and the "ideological effect." In J. Curran et al. (eds.) *Mass Communication and Society*. London: Edward Arnold, pp. 313–348.

Hedebro, G. (1979) *Communication and social change in developing nations: a critical view*. Stockholm: Economic Research Institute.

Hedebro, G. (1982) *Communication and social change in developing nations: a critical view*. Ames: Iowa State University Press.

Hitchcock, Robert (July, 1987) (Anthropologist & Traditional Sector Specialist, Swaziland Manpower Development Project, Community Development Section, Ministry of Agriculture and Co-operatives). Personal Communication. Mbabane, Swaziland.

Hogan, Helen M. (1968) *An ethnography of communication among the Ashanti*. M.A. Thesis. University of Pennsylvania.

Hogben, Lancelot (1969) *The wonderful world of communication*. Garden City, N.Y.: Doubleday.

Hope, Anne and Sally Timmel (1984) *Training for transformation: A handbook for community workers*. Harare: Mambo Press.

Hornik, Robert (1988) *Development communication*. White Plains, New York: Longman.

Hulme, F. E. (1899) *The history, principles, and practice of symbolism in Christian art*, (London: Swan Sennenschien). Quoted in R. Firth, *Symbols*. London: Arnold, 1973.

Huston, P. (1985) Who should talk to whom? *People*, 2: 14–16.

Hyden, G. (1986) African social structure and economic development. In R. J. Berg & J. S. Whitaker (Eds.) *Strategies for African Development* (pp. 52–80). Berkeley: University of California Press.

Hymes, Dell (1972) The use of anthropology. In Dell Hymes (ed.) *Reinventing Anthropology*. New York: Random House, pp. 111–139.

Idiens, D. and K. G.. Ponting (Eds.)(1980) *Textiles of Africa*. Bath: The Pasold Research Fund.

Ingle, H. T. (1972) *Communication media and technology: A look at their role in non-formal education program*. New York: Continuum.

Ito, Youichi (1990) Mass communication theories from a Japanese perspective, *Media, Culture & Society*, 12(4): 423–464.

Jacobson, Howard B. (1969) *A Mass communications dictionary*. New York: Greenwood Press Publishers.

James, Babba (August, 1988) "Use of Improvisational Theatre for the Promotion of Development in Sierra Leone." Oral Presentation. Conference for Communication in Development sponsored by the African Council on Communication Education, Gbarnga, Liberia.

Janis, I. L. (1972) *Victims of groupthink*. Boston: Houghton Mifflin.

Jefkins, F. and F. Ugboajah (1986) *Communication in industrialising countries*. London: Macmillan.

Johnson, S. E. (1932) *Intelligence report on Afaha Obong Clan*, CSO

Papers, Ibadan: National Archives, No. 28242.

Jussawala, M. and D. L. Hughes (1984) Indigenous communication systems and development: a reappraisal. In G. Wang & W. Dissanayake (eds.) *Continuity and Change in Communication Systems: An Asian Perspective*. Norwood, NJ: Ablex, pp. 21–33.

Katz, Elihu and George Wedell (1978) *Broadcasting in the Third World: Promise and performance*. Cambridge, Mass.: Harvard University Press, pp. 102–110.

Keogh, James (1972) *President Nixon and the press*. New York: Funk and Wagnalls.

Kerr, D. (1981) Didactic theatre in Africa, *Harvard Educational Review*, 51(1): 145–155.

Kidd, R. (1984b) Popular theatre and nonformal education in the Third World: Five strands of experience, *International Review of Education* 30(3): 265–287.

Kidd, R. (1983, March). Didactic theater, *Media in Education and Development*, pp. 33–43.K idd, R. (1979) Liberation or domestication: Popular theatre and non-formal education in Africa, *Educational Broadcasting International*, 12: 3–9.

Kidd, R. (1982) *Popular performing arts, non-formal education and social change in the Third World: A bibliography and review essay*. The Hague: Center for the Study of Education in Developing Countries.

Kidd, R. and M. Byram (1977) Popular theatre and development: A Botswana case study, *Convergence (The Journal of the International Council for Adult Education)*, 10(2): 20–30.

Kidd, R. and N. J. Colletta (1980) *Tradition for development, indigenous structures and folk media in non-formal education*. Berlin.

Kivikuru, U. (1989) Communication in transition: The case of Tanzanian villages, *Gazette*, 43(2): 109–130.

Kompaore, P. (August 8, 1989) Personal conversation with Joy Morrison, Ouagadougou, Burkina Faso.

Kreutz, A. (1987) Wearing the message in Niger, *Development Communication Report*, 3: 12.

Langer, Susan (1957) *Philosophy in a new key*. Cambridge, Massachusetts: Harvard University Press.

Laye, C. (1984) *The Guardian of the World* (J. Kirkup, Trans.). New York: Vintage Books.

Legislate minimum age to control marriages. (March, 1990) *People's Daily Graphic*, p.3.

Leis, A. (1979, March) The popular theater and development in Latin America, *International Broadcasting*, pp. 10–13.

Lent, John A. (Sept–Oct 1977) A Third World News Deal? Part One: The Guiding Light, *Index on Censorship*, 16(5): 17–26.

Lerner, Daniel (1958) *The Passing of traditional society.* Glencoe, Ill.: The Free Press.

Luria, A. R. (1976) *Cognitive development: Its cultural and social foundations* (Translated by Martin Lopez-Morillas and Lynn Solotaroff). Cambridge, Mass.: Harvard University Press.

Malamah-Thomas, David (1987) Community theatre with and by the people: The Sierra Leone Experience, *Convergence*, 20(1): 59–68.

Malik, Madhu (1982) *Traditional forms of communication and the mass media in India.* Paris: Unesco.

Mante, J. (November 28, 1989). Child survival and development, *People's Daily Graphic*, p.5.

Manuwuike, E. (1978) *Dysfunctionalism in Afrikan education.* New York: Vantage.

Markus, M. L. (1987) Toward a "critical mass" theory of interactive media: Universal access, interdependence and diffusion, *Communication Research*, 14(5): 491–511.

Marx, Karl (1934) "Thesis on Feuerbach" (Jotted down in Brussels in 1845). In Frederick Engels (ed.) *Ludwig Feuerbach.* New York: International Publishers.

Maxwell, J. (ed.) (1928) *The Gold Coast Handbook, 1928.* 3rd ed. Westminster: Crown Agents for the Colonies for the Government of the Gold Coast.

McIvor, C. (1990) Theatre for Development in Zimbabwe, *D + C*, 3: 29–30.

McLuhan, M. (1964) *Understanding the media: The extensions of man.* New York: McGraw Hill.

McQuail, Denis (1987). *Mass communication theory: An introduction.* 2nd ed. Beverly Hills, CA: Sage.

Melkote, S. (1987) Biases in development support communication, *Gazette*, 40: 39–55.

Merrill, John C. and Ralph L. Lowenstein (1979) *Media, messages, and men.* New York; London: Longman.

Mhlaba, Luke (1991) Local cultures and development in Zimbabwe: The case of Matabeleland. In Preben Kaarsholm (ed.) *Cultural Struggle and Development in Southern Africa.* Harare: Baobab Books, pp. 209–225.

Midgley, J. (ed.) (1986). *Community participation, social development, and the state*. London: Methuen.

Moi, Toril (1985) *Sexual textual politics: Feminist literary theory*. New York: Routledge.

Moore, Henrietta L. (1988) *Feminism and anthropology*. Minneapolis: University of Minnesota Press.

Morrison, J. (1989a) *Good Food Means Good Health* (Consultant's Report on Burkina Faso Nutrition Communication campaign). Washington DC: Academy for Educational Development, (unpublished).

Morrison, Joy (November, 1989b) " '*Fatouma*, the Baby Machine': A case study of forum theatre in a family planning campaign in Burkina Faso." Paper presented at the annual meeting of the African Studies Association.

Morrison, J. (1991) Communication and social change: A case study of forum theater in Burkina Faso. Thesis, Ph.D. (unpublished). University of Iowa, Iowa City.

Mwansa, Dickson (1991) Popular theatre as a means of education: Suggestions from Zambia, *Adult Education and Development*, 19: 127–32.

Mzamae, Mbulelo (1992) "Gender Politics and The Unfolding Culture of Liberation in South Africa." Paper presented at the Fall Workshop on South Africa: Queens University, Kingston, Ontario, Canada.

Nichter, M. and M. Nichter (1986) Health education by appropriate analogy: Using the familiar to explain the new, *Convergence*, 19: 63–71.

Nketia, Kwabena J. H. (1966) *Music in African cultures*. Accra: Institute of African Studies, University of Ghana.

Noah, M. (1980) *Ibibio pioneers in modern Nigerian history*. Uyo: Scholars Press.

Nyerere, Julius K. (1973) *Freedom and development*. New York: OUP.

Obeichina, Emmanuel (1975) *An African popular literature: A study of Onitsha market pamphlets*. London: Cambridge University Press.

Obeichina, Emmanuel (1975) Transition from oral to literary tradition. In *Livre et communication au Nigeria: Essai de vue generaliste, Presence Africaine*, pp. 140–161.

Obeng, Ernest E. (1988) *Ancient Ashanti chieftaincy*. Tema: Ghana Publishing Corporation.

Obeng-Quaidoo, Isaac (1985) Culture and communication research methodologies in Africa: A proposal for change, *Gazette*, 36(2): 109–120.

Obeng-Quaidoo, Isaac (1986) A proposal for new communication research methodologies in Africa, *Africa Media Review*, 1(1): 89–98.

Obeng-Quaidoo, Isaac (1987) New development-oriented models of communication research for Africa: The case for focus group research, *Communicatio Socialis Yearbook*. pp. ....

Odeani, D. (1988) West African mass communications research at major turning point, *Gazette*, 41: 161–183.

Ogan, Christine L. (1982) Development Journalism/Communication: The status of the Concept, *Gazette*, 29: 1–2).

Ofori-Ansa, K. (1983) Africa's search for communication technologies for education: A reflection on problems and prospects, *Development communication Report*, 43: 3, 4–5,13.

Okam, Hilary H. (1978) Oral art forms and literature in Africa. In Christopher C. Mojekwu *et al.* (eds.) *African Society, Culture and Politics: An Introduction to African Studies*. Washington, D.C.: University Press of America, pp. 242–259.

Okita, S. I. O. (1982) Museums as agents for cultural transmission, *Nigeria Magazine*, p.143.

Omibiyi, Mosunbola (1977) Nigerian musical instruments, *Nigeria Magazine*, pp. 122–123.

Ong, W. (1977) *Interfaces of the word*. Ithaca; London: Cornell University Press.

Ong, W. (1982) *Orality and literacy: The technologizing of the World.* London: Methuen.

Opubor, A. (1975) Development communication: A selective bibliography, *Rural Africana*, 27: 127–156.

Organization of African Unity (OAU) (Jan. 5, 1981). *African Charter on Human and People's Rights*. Doc, CAB/LEG/67/3/Rev. (Adopted by OAU Summit in Nairobi, Kenya).

Parmar, S. (1975) *Traditional folk media in India*. New Delhi: Geka Books.

Pask, G. (1976) "An outline theory of media: Education is entertainment." Paper presented to the International Conference on Evaluation and Research in Educational Television and Radio, The Open University, United Kingdom, 9–13 April, 1976.

Pickering, A. K. (1957a) Village drama in Ghana, *International Journal of Adult & Youth Education*, 9(4): 178–83.

Pickering, A. K. (1957b) Village drama in Ghana, *Fundamental and Adult Education*. Paris: Unesco, pp. 178–183.

Pratt, C. (November, 1987) "Communication Research and Development Policy: Agenda Dynamics in an African Setting." Paper presented at the 30th annual African Studies Conference, Denver, Colorado.

Pratt, James (February, 1989) Monthly Report to the Chief of Party. Gbarnga & Monrovia, Liberia (Unpublished).

Pratt, C. B. and J. B. Manheim (1988) Communication research and development policy: Agenda dynamics in an African setting, *Journal of Communication*, 38(3): 75–95.

Pryor, Dennis (1986, September 6) A one-sided marriage of necessity, *The Age Saturday Extra*, p.18.

Pye, L. W. (1963) Modes of traditional, transitional and modern communications systems. In L.W. Pye (ed.) *Communication and Political Development*, Princeton, N.J.: Princeton University Press.

Ranganath, H. K. (1979) *Folk media and communication*. Bangalore: W.Q. Judge.

Ray, Benjamin (1991) *Myth ritual and kingship in Buganda*. New York: Oxford University Press.

Reiser, R. A. and R. M. Gagne (1982) Characteristics of media selection models, *Review of Educational Research*, 52(4): 499–512.

Riley, M. (1990) Indigenous resources in Africa: Unexplored communication potential, *Howard Journal of Communication*, 2(3): 301–314.

Ripley, J. M. (1978) *Communication strategies to aid national development*. Accra: Ghana Universities.

Roberts, P. (1976) The village schoolteacher in Ghana. In J. Goody (ed.) *Changing social structure in Ghana: Essays in the comparative sociology of a new state and an old tradition* (pp. 245–260). London: International African Institute.

Rogers (1983) *Diffusion of innovations* 3rd ed. New York: The Free Press.

Rogers, E. (ed.) (1976a) *Communication and development: Critical perspectives*. Beverly Hills: Sage.

Rogers, Everett M. (April, 1976b) Communication and Development: The Passing of the Dominant Paradigm, *Communication Research*, 3(2).

Ross and Dewdney (1989) Communicating professionally. New York: Allyn & Bacan.

Rothchild, D. and N. Chazan (eds.) (1988) *The precarious balance: State and society in Africa*. Boulder: Westview.

Rusk, Dean (September 25, 1974) Quoted in *The Wall Street Journal*.

Sarpong, P. (1974) *Ghana in retrospect: Some aspects of Ghanaian culture*. Accra: Ghana Publishing Corporation.

Schechner, Richard (1985) *Between theatre and anthropology*. Philadelphia: The University of Pennsylvania Press.

Schramm, Wilbur (1963) Communication development and the

development process. In L.W. Pye (ed.) *Communication and Political Development* (pp.30–57). Princeton, N.J.: Princeton University Press.

Semana, A. R. (1984) Communication, an essential tool in promoting people's participation in rural development. In A-C. Mondjanagni (ed.) *People's Participation in Development in Black Africa*. Paris: Editions Karthala, pp.387–400.

Sieber, Roy (1987) Introduction. In Roy Sieber & Roslyn Adele Walker (eds.) *African Art in the Cycle of Life*. Smithsonian Institution, pp. 84–103.

Siqwana-Ndulo, W. (Fall, 1989) Proverbs: The aesthetics of communication in Africa, *African Studies Center Newsletter*, (UCLA), pp. 20–22.

Smart, N. (1978) The *Densu Times* — self-made literacy, *Development Communication Report*, 21: 1, 3–4.

Sonaike, S. A. (1989) Critical issues in the choice of appropriate communication technology by Third World countries, *Media Asia*, 14(3): 154–161.

Sonaike, S. A. (1987a) Going back to basics: Some ideas on the future directions of Third World communication research, *Gazette*, 40: 79–100.

Sonaike, S. A. (1987) Critical issues in the choice of appropriate communication technology by Third World countries, *Media Asia*, 14(3): 154–161.

Starosta, W. (1974) The use of traditional entertainment forms to stimulate social change, *Quarterly Journal of Speech*, 60: 306–312.

Stevenson, Robert (1988) *Communication, development, and the Third World*. White Plains, N.Y: Longman.

Stewart, A. (1985) Appropriate educational tecnology: Does "appropriateness" have implication for the theoretical framework of educational technology? *Educational Communication Technology Journal*, 33(1): 58–65.

Stone, R. M. (1986) African music performed. In P. Martin & P. O'Meara (Eds.). *Africa* (2nd ed.) (pp. 233–248) Bloomington: Indiana University Press.

Stover, W. J. (1984) *Information technology in the Third World: Can I.T. lead to humane national development?* Boulder, CO: Westview.

Talbot, P. A. (1923) *Life in Southern Nigeria*. London: Frank Cass.

Taylor, J. V. (1950) The development of African drama for education and evangelism. In *The International Review of Missions*, 39, pp. 15–28.

Thomas, P. (October, 1989) Traditional media in a hegemonic culture. Paper presented at the World Association for Christian Communication Congress, Manila.

Traber, M. and K. Nordenstreng (1992) *Few voices, many worlds: Towards a media reform movement*. London: World Association for Christian Communication.

Trevino, L. K., R. H. Lengel and R. L. Daft (1987) Media symbolism, media richness, and media choice in organizations: A symbolic interactionist perspective', *Communication Research*, 14(5): 553–574.

Trudeau, Pierre Elliott (1993) *Memoirs*. Toronto: McClelland and Stewart.

Turner, Victor (1969) *The Ritual Process*. Chicago: Aldine Publishing.

Tunstall, J. (1977) *The media are American*. New York: Columbia University Press.

Turner, Victor (1982) *From Ritual to Theatre*. New York: Performing Arts Journal Publications.

Turner, Victor and Edith Turner (1988) Performing ethnography, *Drama Review*, 26(2): 33–50.

Ugboajah, F. O. (1986) Communication as technology in African rural development, *Africa Media Review*, 1(1): 1–19.

Ugboajah, Frank O. (Autumn, 1985) Developing indigenous communication in Nigeria, *Journal of Communication*, 29(4).

Ugboajah, F. O. (1985c) "Oramedia' in Africa'. In F.O. Ugboajah (ed.) *Mass Communication, Culture and Society in West Africa*. New York: K.G. Saur, pp. 165–176.

Ugboajah, F. O. (1979) Developing indigenous communication, in Nigeria, *Journal of Communication*, 29(4): 40–45.

UNESCO (1975) *Getting the message across*. Paris: UNESCO.

Valbuena, V. T. (1986) *Philippine folk media in development communication*. Singapore: Asian Mass Communication Research and Information Centre.

van Gennep, Arnold (1960) *The rites of passage*. Chicago: University of Chicago Press.

Vieta, K. T. (December, 1989) An unfulfilled task, *West Africa*, pp. 2094–95.

Von Hentig, H. (1986) What are the "media"? What is their rational function and where does it end? *Education*, 34: 43–51.

Wang, G. and W. Dissanayake (eds.) (1984) *Continuity and change in communication systems: An Asian perspective*. Norwood, N.J.: Ablex.

Watzlawick, P., J. Beavin and D. Jackson (1980) The impossibility of not communicating. In John Corner and Jeremy Hawthorn (eds), *Communication studies: An Introductory Reader*. London: Edward

Arnold. First published as *Pragmatics of Communication*, London: Faber, 1976.

White, P. B. (1976) Educational uses of the media: A problem in the social organization of knowledge. Thesis, Ph.D. (unpublished), Syracuse University.

Wilson, D. E. (1975), "The 'drinkard' as a Bunyanian pilgrim", Unpublished Project Report, Department of English, University of Ibadan, Ibadan.

Wilson, Des (1987) Traditional systems of communication in modern African development: An analytical viewpoint, *Africa Media Review*, 1–2: 87–104.

Wilson, Des (1988) Towards integrating traditional and modern communication systems. In Ralph Akinfeleye (ed) *Contemporary Issues in Mass Media for Development and National Security*. Lagos: Unimedia Publications, pp. 67–85.

Yankah, K. (1989) *The proverb in the context of Akan rhetoric*. New York: Peter Lang.

Yankah, K. (1992). The traditional lore in population communication: The case of the Akan in Ghana, *Africa Media Review*, 6: 15–24.

Yankah, K. (1995) *Speaking for the chief: Okyeame and the politics of Akan royal oratory*. Bloomington: Indiana University Press.

Zeeman, T. (1990) Making the NTN relevant in an independent Namibia. In *Some essays on the cultural life in Namibia* (pp. 38–39) Windhoek, Ministry of Education and Culture.

Zimbardo, P. and K. B. Ebbensen (1969) *Influencing attitudes and changing behaviour.* London: Addison-Wesley.

INDEX

abentia, 45–46
adenkum see adziwa group
adinkra ideographs, 245
adziwa, group, 145, 147–148, 206
African Centre for the Training of Performing Artists (ACTPA), 95, 96
African Charter on Human and Peoples' Rights, 27, 28
African media personnel, 159
  training of, 159–160, 165–171
African music, 51, 52
African oral tradition *see* oral tradition
African Research and Educational Puppertry Programme (AREPP), 99
African Symposium Workshop, 96
African Theatre Exchange (ATEX), 96
Aggarwala, Narinder, 30
Agnew, Spiro, 33
Agovi, K. E., 152, 154–155
Akpabot Sam, 40, 42, 47, 54
Alexandre, Pierre, 50
amanfo, 182
Anim, G. T., 31
Ansa, Paul A. V., 4, 9, 18, 23, 159
Ansu-Kyeremeh, 4, 6,7, 9, 11, 17, 20, 21, 185, 186, 187, 190, 241, 242
anthropology, 104, 109
asafo, 145, 146, 153, 156
asafohene, 183
asafotwene, 191
Atelier Theatre Burkinabe (ATB), 131, 134, 136
atumpan, 155
Atwia Story-Telling Group, 80–81
Awa, 8
Bame, K. N., 5, 9, 18, 19, 20, 21–22, 71, 223
Baster community, 92
Benue State Council for Arts and Culture, 88
Boafo, Kwame, 18
Boal, Augusto, 134
Bono, (The), 177–192

political communication system, 178–182
political organization, 183–185
social relationships in, 185–186
utilization of indigeneous communication systems in, 186–192
Botswana, 86–87, 222
Bourgault, Louise M., 5–6, 9, 10–11, 131, 241
Burkina Faso, 130–131, 132, 134–138
Burkinabe *griots,* 132
Busia, K. A., 27
Byram, 87
Chenoff, 107, 169–170
Colleta, Nat, 65, 131
Comic plays *see* Concert-party plays
Communal media, 17
Communication,
  as culture, 231–234
  as industry, 230–231
  diachronic-synchronic view of, 229–230
  games as, 188
  in Akan, 178
  instututional, 58
  oral, 130, 132
  performannce-oriented, 188–189
  problems in Africa, 236–238
  see also indigenous communication systems, traditional modes and instruments of communication
Communication policy for Africa, 229–230
Communication technology,
  modern, 235–236
  traditional 234–235
Concert-party plays, 71–72, 223
  study about, 72–79
  see also folk plays, 'Theatre for development'
Consultative Conference on African Theatre (1983), 95
Cultural performances, 133